Assessment in University Physics Education

IOP Series in Physics Education

The IOP Series in Physics Education aims to provide comprehensive, authoritative and innovative coverage for those that teach physics and related subjects at universities and other higher and further education institutions, and for those involved in physics education research.

Series Editor
Professor Peter Main
King's College London, UK

About the Editor
Peter Main obtained his PhD from the University of Manchester and, after post-docs in Manchester and Helsinki, he joined the University of Nottingham as a Lecturer in Physics in 1979. Following promotions to Reader and Professor, he eventually became Head of the School of Physics and Astronomy. His principal research interests were in quantum fluids and quantum transport in semiconductor and metallic heterostructures. He was also involved in many teaching innovations.

In 2002, he left Nottingham to join the Institute of Physics as Director of Education and Science. In this post, he had overall responsibility for the Institute's work in education at all age levels, research and diversity. Among many projects, he worked closely with Ofqual and awarding bodies on curriculum matters and with government to increase the number of physics teachers. He also initiated several projects improving the diversity of participation in physics.

In 2015, he joined King's College to become Head of Physics; he retains his interest in many projects in physics education and diversity.

About the Series
The IOP Series in Physics Education aims to provide comprehensive, authoritative and innovative coverage for those that teach physics and related subjects at universities and other higher and further education institutions, and for those involved in physics education research.

The series supports evidence-informed professional practice and will cover topics including: assessment methods; feedback; conceptual understanding; problem solving; teaching methods; education technology; pedagogical theory; curriculum design; student engagement; misconceptions; employability; and social aspects of education.

Authors are encouraged to take advantage of electronic publication through the use of colour, animations, video, data files and interactive elements, all of which offer particular benefits in communicating pedagogy.

Do you have an idea for a book you'd like to explore?
We are currently commissioning for the series; if you are interested in writing or editing a book please contact Caroline Mitchell at caroline.mitchell@ioppublishing.org.

Assessment in University Physics Education

Peter Main
King's College London, London, UK

IOP Publishing, Bristol, UK

© IOP Publishing Ltd 2022

All rights reserved. No part of this publication may be reproduced, stored in a retrieval system or transmitted in any form or by any means, electronic, mechanical, photocopying, recording or otherwise, without the prior permission of the publisher, or as expressly permitted by law or under terms agreed with the appropriate rights organization. Multiple copying is permitted in accordance with the terms of licences issued by the Copyright Licensing Agency, the Copyright Clearance Centre and other reproduction rights organizations.

Permission to make use of IOP Publishing content other than as set out above may be sought at permissions@ioppublishing.org.

Peter Main has asserted his right to be identified as the author of this work in accordance with sections 77 and 78 of the Copyright, Designs and Patents Act 1988.

ISBN   978-0-7503-3851-6 (ebook)
ISBN   978-0-7503-3849-3 (print)
ISBN   978-0-7503-3852-3 (myPrint)
ISBN   978-0-7503-3850-9 (mobi)

DOI   10.1088/978-0-7503-3851-6

Version: 20220501

IOP ebooks

British Library Cataloguing-in-Publication Data: A catalogue record for this book is available from the British Library.

Published by IOP Publishing, wholly owned by The Institute of Physics, London

IOP Publishing, Temple Circus, Temple Way, Bristol, BS1 6HG, UK

US Office: IOP Publishing, Inc., 190 North Independence Mall West, Suite 601, Philadelphia, PA 19106, USA

To all the students I have taught.

Contents

Preface		x
Acknowledgement		xi
Author biography		xii

1 Introduction and the purposes of assessment — 1-1

- 1.1 Introduction — 1-1
- 1.2 Purposes of assessment — 1-3
  - 1.2.1 Classifying students: grading — 1-3
  - 1.2.2 External reputation: quality assurance — 1-5
  - 1.2.3 Promoting and measuring learning — 1-7
  - 1.2.4 Benchmarking — 1-10
- 1.3 General remarks — 1-11
- Bibliography — 1-14

2 Methods of assessment — 2-1

- 2.1 Introduction — 2-1
- 2.2 Norm and criterion referencing — 2-2
  - 2.2.1 Norm referencing — 2-3
  - 2.2.2 Criterion referencing — 2-5
  - 2.2.3 Examples of grade descriptors — 2-5
- 2.3 Difficulty and scaling — 2-9
- 2.4 Types of assessment — 2-10
- 2.5 Students with disabilities — 2-16
  - 2.5.1 Specific barriers to learning (e.g. dyslexia or attention deficit hyperactivity disorder etc) — 2-17
  - 2.5.2 Physical disabilities affecting mobility or dexterity — 2-17
  - 2.5.3 Sensory impairments such as deafness or severe loss of sight — 2-18
  - 2.5.4 Social or communication impairments, such as autism — 2-19
  - 2.5.5 Mental health conditions (e.g. anxiety, phobias, anorexia, depression) — 2-19
  - 2.5.6 Long-term health conditions, such as cystic fibrosis — 2-20
  - 2.5.7 Disability and learning outcomes — 2-20
- 2.6 Concluding remarks — 2-22
- Bibliography — 2-22

3	**What are we assessing?**	**3-1**
3.1	Introduction	3-1
3.2	Testing physics programmes	3-2
	3.2.1 Knowledge	3-3
	3.2.2 Mathematical skills	3-4
	3.2.3 Computing	3-5
	3.2.4 Ability to communicate	3-6
	3.2.5 Conceptual analysis	3-10
	3.2.6 Problem solving: logical reasoning and originality	3-12
	3.2.7 Scientific methodology	3-14
	3.2.8 Experimental competence	3-14
	3.2.9 Ethics and professional behaviour	3-14
	3.2.10 Independent learning	3-15
	3.2.11 Team working	3-16
3.3	Assessment of teamwork	3-17
	3.3.1 Team assessment	3-18
	3.3.2 Peer and self-assessment	3-18
	3.3.3 Individual assessment	3-20
	3.3.4 Clarity and transparency	3-20
3.4	Final remarks	3-21
	Bibliography	3-22
4	**Timed examinations**	**4-1**
4.1	Introduction	4-1
4.2	Structure and nomenclature	4-2
	4.2.1 Structure	4-2
	4.2.2 Nomenclature	4-3
	4.2.3 Information for candidates	4-7
4.3	Quality control	4-7
4.4	Marking	4-8
4.5	Analysis of examination papers	4-9
	4.5.1 Year 1 paper	4-9
	4.5.2 A general paper	4-14
	4.5.3 Year 2 paper	4-18
	4.5.4 Option paper	4-22
4.6	General remarks	4-28
	Bibliography	4-30

5	**Experimental and other skills**	**5-1**
5.1	Investigative work	5-1
5.2	Projects	5-5
	5.2.1 Types of projects	5-6
	5.2.2 Principles of assessing extended projects	5-8
5.3	Types of assessment	5-9
5.4	Examples of assessment criteria	5-12
	5.4.1 Logs and performance	5-13
	5.4.2 Written and oral presentations	5-25
	5.4.3 Other types of assessment	5-31
5.5	General remarks	5-35
	Bibliography	5-37
6	**Summaries and suggestions**	**6-1**
6.1	Introduction	6-1
6.2	Emergent issues	6-2
6.3	Suggestions for improvement	6-7
	6.3.1 Quality assurance and assessment	6-8
	6.3.2 Norm versus criterion referencing	6-10
	6.3.3 Designing assessment	6-11
	6.3.4 Variety of assessments	6-13
	6.3.5 Combining marks	6-16
	6.3.6 Combining marks within modules	6-18
	6.3.7 Consistency	6-19
	6.3.8 Instigating change	6-20
6.4	Summary	6-22
	Bibliography	6-22

Preface

This book is about assessment in university degree programmes in physics, although much of the content would be of interest to assessors in other disciplines. It is very much a practical book, providing advice for academics involved in the process of assessment. Although there are some references to educational research, this is a book written from the perspective of experience and observation.

There are three main purposes to the book. First, it serves as a practical guide for new members of staff who are inexperienced in teaching and assessment. Second, it critiques existing forms of assessment, so that more experienced examiners and programme leaders are encouraged to think more deeply about the nature of their assessments. Finally, it addresses broader issues within the assessment system which have consequences for senior administrators, quality assurance experts and those interested in university standards.

Above all, the book sets out to encourage examiners to think about the purposes of assessment and whether the approaches they use satisfy those purposes. If it achieves that goal, the writing of it will be time well spent.

Acknowledgement

The book is based on my experience as a university teacher and examiner. It follows that my fellow academics, in my own universities and those where I have had the pleasure of being an external examiner, have had a profound effect on my thinking. There are far too many to mention but I am grateful for all their comments over several decades. In the creation of this book, I have been grateful for helpful exchanges with many people; among the more prominent are, in alphabetical order, George Booth, Carl Brown, Mark Fromhold, Jeff Grube, Helen Heath, Sally Jordan, Samjid Mannan, David Mowbray, Eva Philippaki, David Sands, Alison Voice and Furqaan Yusaf.

I am grateful to the following universities for allowing me to use examples of assessment material: University of Bristol, King's College London, University of Leeds, University of Nottingham and the University of Sheffield.

This is a book in an IOP Series of which I am Series Editor. I resist the unworthy temptation to thank myself but I am indebted to the enthusiasm and support of Caroline Mitchell and the technical help from Robbie Trevelyan and Calum Sims at IOP Publishing.

Finally, I would like to thank Vesna, who made a number of very helpful suggestions from a non-physics perspective. She also persuaded me that I would not survive retirement unless I had a project to pursue. She was, undoubtedly, correct.

Author biography

Peter Main

Peter Main graduated from the University of Birmingham in 1973 and obtained his PhD from the University of Manchester in 1976. After a post-doc in Manchester and a NATO Fellowship in Helsinki, he joined the University of Nottingham as Lecturer in 1979, later obtaining a Chair and becoming Head of the School of Physics and Astronomy. His research included work on liquid helium and the quantum properties of metallic and semiconducting nanostructures. He was also closely involved with several educational innovations, one of which was the development in the early 1990s of the Nottingham MSci, which had a heavy focus on student-centred activity and included one of the first examples of group projects based in industry.

In 2002 he left Nottingham to take on a new role of Director of Education and Science at the Institute of Physics, with responsibility for research matters, education and diversity. His work associated with higher education included a leading role in the establishment of Project Juno, a scheme to promote gender equality in physics departments, and the creation of the Higher Education Group, bringing together academics who had a strong interest in education.

In 2015 he returned to academia as Head of Physics at King's College London where he stayed until his retirement in the summer of 2020.

IOP Publishing

Assessment in University Physics Education

Peter Main

Chapter 1

Introduction and the purposes of assessment

1.1 Introduction

Across the whole of the formal learning spectrum, assessment is inextricably linked to education and, for many students, the reason for learning is to achieve some sort of a qualification. That qualification may be general, such as a Baccalaureate or a set of A-levels, or it may be something quite specific, such as a formal requirement for professional practice or even the right to drive a motor vehicle. In many countries, and certainly within the UK, the nature and form of the assessment has had an increasing influence on the learning process. In England, where there are substantial fees for university tuition, the student attitude has shifted away from obtaining an education towards paying for a qualification. Further, university finances depend crucially on those student fees and that combination has put a strain on the nature and integrity of the assessment process. A feature of the last couple of decades has been the inexorable grade inflation in universities to the extent that many students now consider anything but a first-class honours degree a disappointment.

As in other aspects of life, the Covid-19 crisis has had a profound effect on universities. One major issue was that most universities were forced to abandon traditional, timed examinations in favour of some sort of open-book assessment. In science disciplines in particular, this rapid rearrangement caused huge disruption and, in many cases, led to a reappraisal of the form and nature of the assessment process. While the Covid crisis continues the matter is unlikely to be resolved fully, but the situation does offer an opportunity to rethink assessment and perhaps move towards a process that is more fit for purpose.

In this book I consider assessment in the context of a university department of physics. However, many of the points considered would be valid in any educational environment, including schools, and certainly in other university disciplines. My thoughts are based on more than 40 years' experience of assessment, including roles as a university teacher, an external examiner and extensive involvement in quality assurance procedures. Anyone seeking a comprehensive overview of research in this area is referred to a recent review by McConlogue [1], which I found very useful.

The aim of the book is to provide a context for practitioners to review and, I hope, improve their assessment processes; it is purposely written to provide concrete examples rather than rely on abstract principles so that academics may relate it to their own teaching. The various chapters look at the purposes and types of assessment, how they match what we are trying to achieve with our teaching (including an analysis of current assessments) plus some discussion points on how we might go forward. There are also suggestions for how the government might reconsider its process for assessing the quality of university education in the UK. I devote one chapter to the assessment of practical work, which is often seen as one of the most valuable parts of a degree programme, while posing special challenges for assessment.

Although there is a great deal wrong with how various government agencies have tried to assess and thereby improve the quality of university teaching, one undeniable consequence is that most universities now encourage academics to think more deeply about their teaching and, by association, the corresponding assessment. However, my experience is that, while the majority of staff do care a great deal about their teaching, the assessment of theory modules/courses has varied little from the traditional unseen examination. I hope this book encourages people to think more carefully about what they are trying to achieve.

Another factor that has had a profound impact on the nature of assessment is the transparency of the process. From when I was an undergraduate in the 1970s, right through to my early years as an academic, broadly speaking students were given no quantitative information about their various assessments, with the possible exception of those associated with laboratory work. There were no mark transcripts, just the bare degree classification. Nowadays, not only do students know the marks for every element of assessment, in many universities they also have access to their marked examination papers. In general, this shift is a massive improvement but, along with other factors discussed later, it has tended to shift the focus away from education towards assessment,

As a final point, I want to point out a peculiarity of the system in most, but not all, universities. It is typically the case that the same person teaches the material, constructs the assessment and carries out the marking. While there are checks and balances for the latter two stages, at least, and no one can deny the consistency of the process, such an arrangement is not common in other assessment environments. One's driving instructor does not act as examiner for the driving test and schoolchildren take national examinations, not those set and marked by the teacher. In some universities, I have known promotion applications refer to the high marks gained by students in modules taught by the applicant as an indicator of excellence. It requires little imagination to see the perils of that approach.

1.2 Purposes of assessment

A useful starting point in any discussion of assessment, or indeed any human activity, is to ask oneself why one is doing it. I recommend asking one of your colleagues, perhaps one who follows the more traditional approach to assessment, what they believe the purpose is. You are likely to obtain a response along the lines of: 'to find out what the students know/understand about the subject'. It is also common for people to believe that setting examination questions is a relatively exact science and that any difference in average marks between modules is due to differences in the collective student aptitude for those modules. One former colleague took this to its logical extreme by setting the same questions every year, in his words to guarantee that students were being measured against the same criteria. One year he changed a question and the students complained at the unfairness of the examination.

In reality, there are many different purposes for assessment, some overlapping and some almost in opposition. For each of the cases below, I try to explain the purpose, where relevant, to show how it might relate to the teaching and learning on a degree programme in physics and to comment on any hidden assumptions being made. The purposes are in no particular order and certainly not in order of importance.

1.2.1 Classifying students: grading

The grading of students does not contribute in any meaningful way to their education. However, for many students, the degree classification that they obtain is seen as the most important outcome of their period at university. And no wonder, when recruiters report that they do not consider applicants without at least an upper second-class degree [2]. In addition, the most popular next step for physics graduates is some sort of further study, often a PhD programme and post-graduate admissions tutors, particularly those in research-intensive institutions with a high demand for places, are increasingly looking for students holding first-class degrees.

The system requires grading of graduates. However, the more one reflects on that process, the more unsatisfactory it appears. In the UK, most universities use a system of degree classifications comprising: BSc (Ordinary), BSc (Honours) pass, BSc (Honours) third class, BSc (Honours) second class upper division—usually known as 2(i); BSc (Honours) second class lower division, or 2(ii); and BSc (Honours) first class. This arcane nomenclature has historical origins [3] and little contemporary significance and there has been much debate in recent years about whether it should be changed, possibly to a grade point average (GPA) system of the type common, for example, in the United States.

There is usually some sort of algorithm that translates module marks or grades into these degree classifications. These algorithms can be astonishingly complex, as if to instill greater confidence, but in practice those complexities do little to address the issues inherent in the process. The first issue is the sheer arbitrariness of drawing lines at various points in the mark distribution. As an example from within the UK system, a first class degree is usually associated with an overall mark of 70% from whatever algorithm is used to determine that mark—the algorithms differ from one university to another. The next class down, 2(i), stretches from 60% to 69.9% although marks above 69% are usually rounded up to first-class. Even if one accepts the validity of the marks, it is clear that the achievement of the bottom first-class student is closer to that of the top 2(i) student than to that of the person who has the highest marks in the class.

A GPA system offers a continuum, removing these arbitrary lines, but I note that for, say, recruitment to PhD programmes, a threshold GPA is often required and that means a line is drawn and similar comments would apply. Nonetheless, GPAs do reflect better the spread of student achievement.

The second issue around degree grading is the lack of comparability between different institutions and even different departments within the same institutions. Even if we swallow for the moment any doubts we might have about what the classifications might mean in terms of a student's aptitude for work or further study, does anyone within the system believe that a first-class Honours degree from an internationally leading Russell Group university means the same thing as one from a small post-1992 institution? (In 1992 the UK higher education system was changed radically and the number of institutions with the right to confer degrees was expanded dramatically.) This is not an elitist point and I have purposely avoided implying that one degree is 'better' than the other, rather I am saying that they are not equivalent (there is an interesting discussion paper by Andrew Hindmarsh [4]). However, several principals and vice-chancellors, with commendable solidarity but somewhat less plausibility, have publicly defended this equivalence against politicians trying to incorporate more competition into the university sector. A GPA system does not escape this comparability issue, either.

In the UK physics community we are extremely fortunate in having a degree accreditation system, administered by the Institute of Physics (IOP) but owned by the academics whose programmes are being accredited [5]. The process involves submission of relevant paperwork, including examples of assessment, and a visit by a team of physicists from other institutions. The process is highly beneficial for the department under accreditation and, while the scheme is not very prescriptive in terms of curriculum content, it does set a threshold standard for the level at which the physics is taught and assessed as well other parameters such as library provision, project work and pastoral support. Historically, the IOP has also worked closely with the Quality Assurance Agency to develop a Subject Benchmark Statement for Physics [6], which I would strongly recommend to anyone unfamiliar with it. Unsurprisingly, the accreditation criteria and the Benchmark Statement are consistent and complementary. However, while they both define a threshold standard, neither attempts to define or compare degree classifications.

In defending the equivalence of degree classifications between institutions, people often point to the presence of external examiners, who are experienced academics from a comparable institution and who oversee the entire examination process, from the setting of the assessments to the degree award. I have myself taken on this role in several institutions. While such examiners generally do their work conscientiously and act as a positive influence in improving both teaching and assessment, it is virtually unknown for an external examiner to call for a substantial redrawing of the grade boundaries. Their role is more to ensure that assessments and processes are fair and consistent than to provide a detailed comparison of standards. Further, it has been documented [7] that universities tend to select their external examiners from other universities 'like them', again militating against any robust comparison.

1.2.2 External reputation: quality assurance

Strictly speaking this is not a purpose of assessment *per se* but an application for which assessment is used. However, successive UK governments have tried to introduce more competition into education at all levels and, since this policy has had a profound effect on how assessment operates in universities, it forms part of the remit of this book. In addition, many newspapers and websites provide ranked lists of the 'best' universities for each subject and these lists are taken seriously by potential students and their advisors (see for example [8]). At least one of these newspapers uses 'value added' as a parameter of quality in determining the alleged quality of the provision—essentially this boils down to giving credit for higher degree grades.

Within the university system, in recent years there have been two major drivers for (alleged) quality assurance, although they are not independent. The first is the National Student Survey (NSS), a survey of all graduating students which asks them to rank their institution and programme according to questions around the quality of teaching, feedback, library provision, teaching accommodation as well as an overall satisfaction rating. There is also an opportunity for students to add free-text comments.

The second driver is the so-called Teaching Excellence Framework (TEF) [9], which is administered by the Office for Students (OfS). At the time of writing, the TEF is still not fully developed—the OfS has commissioned an independent review—although many universities do have awards (bronze, silver or gold) at both institution and subject levels. The intention is that the TEF ranking of a university will feed into its funding, a feature which has ensured that the universities take it very seriously indeed. In its early versions, the TEF has attracted much criticism from academics, largely because it seemed to have little to do with teaching excellence.

The four main parameters of the TEF are, in broad terms:
1. Input from the NSS.
2. Employability of graduates.
3. Progression rates from year 1 to year 2.
4. A narrative which discusses the numerical data.

The NSS is not a good measure of teaching quality. There is much anecdotal evidence that innovative teaching and assessment, particularly the types that challenge students, are unpopular and can lead to low scores. Further, students have only their own experience to draw on and simply do not know if other institutions do things better. As a final point, it is common for the graduating cohort to provide just a few dozen survey returns for a programme, which leads to large statistical fluctuations in the numerical scores. These fluctuations are random but the nature of the TEF required departments to explain, say, why the overall satisfaction rating had dropped from 90% to 83%.

The employability of graduates is also not a good measure of the quality of teaching and assessment, particularly when the data are derived from the university's own returns and within a short time of graduation. The employability of students will depend heavily on the subject they have taken—most physics departments are in the upper quartile of employability—and on the reputation of the institution [10].

Progression rates from year 1 to year 2 may seem to offer a reasonable measure of teaching quality but a few moments' thought reveals any number of ways the system can be played. The most obvious is to set a low hurdle at the end of year 1. The measure also offers a strong steer to universities towards a 'safe' admissions policy, all the more so since the progression rate criterion makes no allowance for students who were ill or had other good reasons to delay their progression.

The narrative input allows departments and universities to provide some context for the returned data but the overall pressure of the TEF is on assessment standards. The link to the progression rate is obvious but it is also true that, for a given institution, the best way to improve the employability of your graduates is to give them better degrees. Similarly, if you want happy students for the NSS, give them high marks.

While this is a simplistic view and there are many academics who are fierce defenders of standards, nonetheless, these various pressures have coincided with a dramatic grade inflation in degree results over the last decade. Table 1.1 shows figures from UK universities in three indicative years over the last decade.

The proportion of degrees that are first class has almost doubled in 7 years, while those receiving degrees classed below upper second has dropped by more than a third in the same period. While there are those who suggest that this rise really does represent a massive improvement in student attainment, I am not aware of any evidence to support that view and it is more common to see the inflation as a change in assessment standards driven by the external pressures exerted by the government. Both students and universities benefit from grade inflation, in the short term at least,

Table 1.1. Degree classifications in three indicative years. (Source: Office for Students [11]).

Degree classification	2010/11		2016/17		2017/18	
	Number	%	Number	%	Number	%
First	34,910	15.7	68,990	27.2	75,840	29.3
Upper second	114,075	51.3	128,550	50.7	128,800	49.7
Other	73,445	33.0	56,030	22.1	54,545	21.0
Total	222,430	100	253,570	100	259,185	100

and there is virtually nothing in the system to provide an incentive against that trend. The government has bridled against this trend [12] and threatened sanctions but it might do better to look at the regulatory environment it has created and its effect on assessment. This point will be discussed further in the final chapter. The issue of standards will be discussed in detail in chapter 2.

1.2.3 Promoting and measuring learning

Whereas most students see assessment as a route to a qualification, many academics would cite the promotion of learning as its primary purpose. While it is true that learning usually requires an incentive, that does not necessarily imply an examination. Rather late in life, I am trying to learn a musical instrument with no more an incentive than to make a few tuneful noises. People learn languages to live abroad or simply because they want to. But assessment is so much a feature of formal education that, inevitably, how students approach their learning is heavily influenced by the nature of the assessment. A colleague in my most recent university went to great trouble to set up sessions involving student-based learning to supplement his lectures. These sessions followed immediately after the lectures so students did not even need to make an extra journey. Nonetheless, as soon as the class knew that the sessions were not part of the assessment, half of them walked out at the end of the formal lecture.

At this point, it is useful to draw a distinction between *formative* and *summative* assessment, although that distinction is not as sharp as is often presented. Briefly, summative assessment is the overall assessment of student learning in the module; if you wish, the assessment that counts towards the marks or grades awarded for that course or module. Formative assessment is more diagnostic; its purpose is to help students determine their progress, how well they are coping with the material etc. It can also be helpful for the teacher to determine which topics or elements have been grasped by the students and which require reinforcement.

There are a number of reasons why summative and formative assessments are not so clearly delineated. Formative assessment is only effective if students actually take the assessment and also take it seriously; there is also a requirement for timely feedback. In teaching environments that require student activity, such as computer programming workshops, projects or laboratory work, it is relatively easy to embed formative assessment into the teaching and such methods will be discussed in more detail in chapter 5. In teaching where the summative assessment is a final examination paper, it is less straightforward. Typically, students are given problem sheets associated with the material being taught and there are problem classes with tutorial support to provide rapid feedback. However, in the UK, despite the dubious educational value of the procedure, in most universities teachers are expected to provide full solutions to any problems handed out. Consequently, many students do not attempt the problems—the problem classes are usually sparsely populated—but simply wait for the solutions to appear, thereby eschewing the real benefit of the exercise, which is to determine how well they can apply the material.

One method of ensuring that formative assessment plays a useful role is to give it a small summative element. A typical example is to have regular course work, which is handed in, marked, often by research students, and returned with feedback. Assigning even 10% of the marks for the whole course, typically 1% per weekly set of questions, is sufficient to ensure that the vast majority of students submit the work. It seems that, while students are not keen on assessments with no marks, even a small number is sufficient to persuade them to participate and that participation provides a material improvement in their learning. Such arrangements do allow the opportunity for plagiarism but, while that cannot be eliminated entirely, there are methods of discouraging cheating.

As a brief aside, on a point that will be considered further in later chapters, an important skill in life is the ability to work with other people and use them as a resource. When we have a problem, if we cannot find an immediate solution, the best approach is to find someone who can. And yet we rarely teach this skill. One of the benefits of regular, assessed course work is that students look at it together in groups and actually talk to each other about physics. Maybe the work submitted is not, therefore, entirely a student's own but they will have benefited hugely from the conversations. Wholesale copying, of course, is never a good thing.

If an assessment is to be successful in driving learning, it is essential that what is being assessed is linked to the learning objectives of the teaching. For those that bridle at the mere mention of learning objectives, all this means is that we should be assessing whatever knowledge and skills we want the students to have as a result of the teaching. In some cases, the matching is easy to achieve. For example, in an introductory computing course, the students will usually be learning some high-level language and how to apply it to a range of situations. In such cases, the assessment will involve students writing programs and solving straightforward problems. A formal, written examination could be used as an assessment—it would be straightforward and easy to mark—but it would not be an appropriate measure of the central purpose of the teaching. Similar remarks apply to the teaching of experimental and transferable skills. The link between assessment and learning objectives will be explored further in chapter 3.

By far the most common form of assessment of lecture-based teaching is the timed, unseen, formal written examination. The word 'unseen' means that the students do not know in advance the detailed topics to be covered nor have access to the teaching material, text-books or the Internet. Such examinations are usually taken by all students at the same time, typically at the same venue, and have a fixed duration of 2 or 3 h, although both of those conditions may be relaxed for students with disabilities (see chapter 3 for a discussion on the assessment of students with disabilities). The relative qualities of this type of assessment will be considered in detail in the next chapter; here, the question is: what effect does it have in promoting student learning? My own experience as an examiner, vetting questions set by colleagues and as an external examiner in other institutions, indicates that the principal influence of traditional examinations on student behaviour is to promote rote learning [13]. In chapter 4 there is some analysis of physics examination papers and it is clear that a student depending solely on rote learning will be able to achieve marks consistent with a good degree and, in many cases, a first-class one. A further observation is that generally, the more advanced the material, the more likely the examination will be skewed towards rote learning. There is good reason for this, of course: the relatively short length of the examination and the usual restriction that a simple calculator is the only permitted technological aid do not sit well with any attempt to require students to demonstrate the ability to undertake extended calculations. And there is undoubtedly some virtue in rote learning—few would argue that knowing by heart Maxwell's equations, the Lorentz transformations or Schrödinger's equation, for example, is not key to understanding them.

But how do students respond to this type of assessment? Given the emphasis on learning material by heart, it becomes essential for students to obtain a definitive collection of material upon which to draw. As mentioned above, in most UK universities, it has become obligatory for lecturers to supply a full set of lecture notes; further, it is expected that worked solutions are supplied to any problems handed out. In addition, in many universities, old examination papers are published, sometimes with model answers provided too. Indeed, a few years ago, many physics departments in UK universities were served with a freedom of information request to supply all questions and solutions to their examinations over the previous few years. In such an environment, rote learning is the safest way to ensure a decent mark and even the most able students, with deep understanding of the subject, would be foolish not to place a large emphasis on it.

A further point is that, with the fear of student criticism via the NSS, most UK universities would allow a student appeal if an assessment diverged from the material provided by the lecturer. It is a common complaint from the students that the assessment was not related to the teaching; on investigation, what they usually mean is that they have been asked to apply some of the material instead of simply restating it.

Another aspect of unseen examinations is that, although most questions do include some sort of problem to be solved, in many topics, there are only a few examples that are suitable for this format. Inevitably, questions repeat every few years. As a consequence, almost every student tries to spot questions that are likely to appear in the paper they are taking and they ensure they know the corresponding answers. And while tutors and advisors highlight the dangers of the exercise, even a cursory inspection of a sequence of examination papers indicates that the students have a point.

I finish the section with an anecdote that illustrates some of the points above. I became aware of a student appeal, not in physics but in a cognate subject. A question had asked for a derivation that had been part of the lecture notes distributed to students. However, the lecturer had made an error in the derivation as sent out. In the examination, the student wrote out the derivation as circulated, complete with error, which was subsequently marked as incorrect. The appeal was based on the idea that it was not the student's fault that there was a mistake and that they had reproduced what was asked of them. I am pleased to say that the appeal was rejected.

Assessment to drive learning is its most important academic purpose and should therefore be linked directly to what the teaching is trying to achieve. The extent to which that is the case will be discussed later, together with some suggestions on how that linkage might be improved.

1.2.4 Benchmarking

Figure 1.1 shows the original version of a cartoon that is routinely shown to inject humour into talks on assessment [14]. The intention (I think) is to illustrate how we are all different and that it is, in general, unfair to ask each candidate to carry out the same task. In other words, it illustrates the point that treating everyone the same does not mean giving everyone an equality of opportunity. I have used it myself but, having raised a few smiles, I follow up by asking the audience how they think this analysis applies if the course under assessment happens to be one on tree climbing.

Many assessments are concerned with benchmarking, that is, assessing whether someone, or even something, has reached a threshold standard. In the case of the somethings, it might be, say, a fire-safety standard or the maximum load a cable can support without snapping. For humans, examples include driving tests of various sorts, musical grades and qualification standards for athletics events. There are also benchmark standards in the medical professions for such techniques as drug delivery.

Benchmark assessments are usually, but not always, related to some sort of technical ability, the competence to achieve a particular outcome usually to some specific, tightly-defined criteria. That does not mean that such assessments are wholly objective—the driving test is a good example where the examiner has to make judgements on what is a complex set of skills. Another feature is that their specificity means that it is not always possible to devise an alternative assessment.

Figure 1.1. The nature of assessment. The teacher is saying: 'In the interests of a fair assessment, the task for everyone is the same: climb that tree.' (Every effort has been made to trace copyright holders and to obtain their permission for the use of copyright material within this resource.)

The goldfish in the cartoon might be better advised to choose a different course from the one on climbing trees.

In the context of a physics programme, there are many examples where students are expected to have achieved a particular skill or competence. The most obvious examples occur in laboratory training and computer programming. In the former case, the ability to handle a specific piece of equipment might be relevant whereas, for the latter, the successful application to solve a differential equation could be a necessary skill. However, despite the numerous examples, it is rare for the structure of an assessment to include a benchmark element.

Why is this? The essence of a benchmark assessment is that it is a simple pass/fail. Due to the need for classification and the almost mystical significance given to the accumulation of marks and/or grades into a final, overall mark, digital pass/fail outcomes are hard to accommodate. It is expected that all assessments provide a numerical mark or indicative grade, which in turn requires a distribution of marks, preferable over a broad range, to facilitate this classification.

In fact, the idea of benchmarking, in the sense of the student achieving some sort of minimum standard, could be applied much more widely across the academic spectrum, even in lecture-based modules. In most modules, there are competences that students are expected to have acquired. They might be the ability to solve certain types of equations, or to be able to apply equations to simple physical situations. However, although such outcomes are tested, they are almost never done so in terms of a benchmark pass/fail. Some suggestions on how that might be possible are presented in chapter 6.

1.3 General remarks

Having looked at the purposes of assessment and before moving on to consider the different types, it is helpful to consider the basic principles underlying the fairness of any form of assessment. Different people describe these principles in different ways [15] but I use the following.

Any assessment should be:
- Valid.
- Appropriate and transparent.
- Authentic.
- Accurate and consistent.
- Achievable.

In order to be *valid*, an assessment must be related to whatever the students are supposed to be able to do either at the end of the module/course or at some intermediate point. While that may seem obvious, as discussed above, in many cases where the sole assessment is an unseen, timed, written examination, it is not always clear that the skills and knowledge required to do well in the test are necessarily those that the teacher would prefer.

Appropriateness and *transparency* cover a range of issues. To begin with the latter, the following must be clear to students: what the assessments are, i.e. type(s) and what is covered; when they occur, including any deadlines; and an indication of the level required to achieve whatever grades there are. The last point is actually difficult to achieve. It is common, for example, to supply marking criteria but, as will be discussed in more detail in chapter 5, unless these are very detailed and specific to the particular course, they are far from definitive. In practice, therefore, most students obtain a feel for the level by the simple expediency of looking at old exam papers. While that is sensible and pragmatic, it does lead to a circularity in the sense that the transparency of the process lies in its continuity; very often, a change of teacher for a module leads to a lower average mark as this continuity is lost.

The *appropriateness* of an assessment requires that it is related to the material taught, is comprehensive and is applicable to all the students. The last point can be a delicate one in the context of disabled students. The assessment must also be at an appropriate level—it would not be fair, for instance, to have an examination asking for solutions far more challenging than those students have met within the module.

The *authenticity* of an assessment mainly relates to the confidence we have that it is related to the student's learning or their ability to perform a particular task, depending on the purpose of the assessment. At the simplest level, we can ask the question: is this the work of the student? For unseen examinations this type of authenticity is rarely a problem, but for course work and group work of any type it can lead to issues. For example, there can be a tension between the need to produce grades for individual students and a learning objective that students know how to work in a team. Another common strain on authenticity is when a project student is based in a research laboratory, often using a post graduate student as a mentor. These tensions will be explored further in chapter 5.

Another element of authenticity is that the assessment needs to be relevant to the content of what is being taught, not only in terms of topic but also in terms of skills. As discussed above, assessing a module relating to computer programming by means of an unseen, handwritten examination may raise questions of authenticity, as the outcome of the module is surely about the students being able to write working programs.

In the UK, most degree programmes take a modular form as do the corresponding assessments. However, it is possible, indeed desirable, that there might be learning objectives that apply at the programme level. To some extent, these can be distributed across the modules but the nature of physics is such that some ability to apply knowledge and skills from different areas at a synoptic level is a defining feature of the subject. It follows that a truly authentic assessment at the programme level should go beyond a simple agglomeration of the individual module assessments.

In one sense, of course, no assessment, and certainly no traditional examination, is truly authentic in the sense of measuring what it is to be a physicist. In the end, an assessment measures how well a student performs that assessment and nothing more, and there are few jobs that require one to sit examinations. This is one reason why it is important to choose assessments that drive effective student learning; the assessment is a tool not the goal.

Accuracy and *consistency* are key elements of fairness. At the most basic level, students should not be asked to do impossible things thus, in physics more than most subjects, it is essential to ensure that assessments are free from errors. Most institutions have elaborate procedures for checking the accuracy of examination papers. Accuracy is less of an issue in continually assessed work.

An advantage of timed, written examinations is that all students are treated consistently in that they all do the same paper and are almost always marked by one person. Consistency, however, is much more of an issue when the exercise is individual to the student. Some sort of project is a ubiquitous feature of physics degree programmes and it is common for the project marks to form a substantial fraction of the final year's total, contributing a great deal to the degree classification. In educational terms, a project is a highly desirable component in the education of a physicist but can lead to real issues with consistency of treatment. In reference to the purposes of assessment listed above, projects fit perfectly with the goal of driving learning but pose problems with grading and classification.

Finally, any assessment must be *achievable*. A typical student must reasonably be able to cope with it both in terms of level and in terms of the time available. In practice, achievability follows from historical precedence and it is noticeable that problems that do arise are usually associated with inexperienced teachers. One learns the capability of a class from experience—a deceptively simple remark but one which indicates an important underlying feature of assessment that is explored in more detail in the next chapter.

A common observation in UK universities, and largely a consequence of the modular approach, is that students are over-assessed [16] and that this, in turn, leads to unacceptable levels of stress on students and concerns about their mental well-being. While it is certainly true that mental health issues reported by students have increased markedly in number over recent years, it is less clear that the situation would always be improved by having fewer assessments. One reason is that fewer assessments imply increased weight for those that do exist and, consequently, an increased fear of doing badly. In other words, doing badly in an assessment that comprises 10% of the total mark is not so much of a disaster as it would be if the weighting were 90%.

Another reason for not reducing the number of assessments simply for the sake of doing so is that there is considerable educational benefit in having regular and frequent tests. Not only does that ensure that students keep on top of the material but it also allows the opportunity to provide feedback to both staff and students on how things are going. While one might argue that the assessment does not have to be summative for that to be true, experience shows that many students do not engage with assessment unless it counts in some way.

Achievability must also apply to the assessors. There is little point in creating the perfect test of learning outcomes if it takes half the staff in a department several weeks to mark it.

To conclude the chapter, it is evident that assessment serves a number of purposes in the education system. Those purposes are not always in harmony and some of the tensions between them will be featured in later chapters. The most important of the purposes is undoubtedly to promote education and learning, followed by society's apparent need for universities to classify their graduates. The guiding principle, however, should be that assessment is the servant of learning. We do not teach to assess, we teach so that students learn and that learning should not be compromised by the spurious requirements of assessment.

Bibliography

[1] McConlogue T 2020 *Assessment and Feedback in Higher Education* (London: UCL Press) ch 8
[2] Coughlan S 2010 Three quarters of employers 'require 2:1 degree' *BBC News* (online) www.bbc.co.uk/news/10506798
[3] Alderman G 2013 Tear up the class system *The Guardian* (online) https://theguardian.com/education/2003/oct/14/administration.highereducation
[4] Hindmarsh A 2018 Are degree standards the same at all universities? *HEPI* www.hepi.ac.uk/2018/07/02/degree-standards-universities/
[5] The Institute of Physics 2011 The Physics Degree *IOP* www.iop.org/sites/default/files/2019-10/the-physics-degree.pdf
[6] Quality Assurance Association 2019 Subject Benchmark Statement: Physics, Astronomy and Astrophysics *QAA* www.qaa.ac.uk/docs/qaa/subject-benchmark-statements/subject-benchmark-statement-physics-astronomy-and-astrophysics.pdf
[7] Higher Education Academy, corp creators 2015 A review of external examining arrangements across the UK: report to the UK higher education funding bodies by the Higher Education Academy: June 2015 *HEFCE* https://dera.ioe.ac.uk//23541/
[8] The Guardian 2021 The Guardian UK University Rankings *The Guardian* (online) www.theguardian.com/education/ng-interactive/2021/sep/11/the-best-uk-universities-2022-rankings
[9] Office for Students 2020 About the TEF *Office for Students* https://officeforstudents.org.uk/advice-and-guidance/teaching/about-the-tef/
[10] Brown P and Hesketh A 2004 *The Mismanagement of Talent: Employability and Jobs in the Knowledge Economy* (Oxford: Oxford University Press)
[11] Office for Students 2020 *Analysis of Degree Classifications Over Time: Changes in Graduate Attainment from 2010–11 to 2018–19* OfS 2020.52 Office for Students
[12] Hinds D 2019 Universities told to end grade inflation *Gov.UK* https://gov.uk/government/news/universities-told-to-end-grade-inflation
[13] Chuderski A 2016 Time pressure prevents relational learning *Learn. Individ. Differ* **49** 361–5
[14] Traxler H 1983 Chancengleichheit *Schul-Spott: Karikaturen aus 2500 Jahren Pädagogik* ed M Klant (Hannover: Fackelträger)
[15] Imperial College London What are the qualities of good assessment? https://imperial.ac.uk/staff/educational-development/teaching-toolkit/assessment-and-feedback/good-assessment/
[16] Harland T, McLean A, Wass R, Miller E and Sim K N 2015 An assessment arms race and its fallout; high-stakes grading and the case for slow scholarship *Assess. Evaluat. Higher Educ.* **40** 528–41

IOP Publishing

Assessment in University Physics Education

Peter Main

Chapter 2

Methods of assessment

2.1 Introduction

Chapter 1 described the various purposes that are related to assessment. In this chapter I consider the different types of assessment that might be relevant for a physics or physics-related degree programme. For some of the purposes it does not really matter what methodology is used. For grading, anything that provides a range of marks across a reasonably broad spectrum will do. In contrast, a benchmarking approach may require a simple pass/fail, which, if coupled with the ability to retake the test, may have 100% pass rates. However, if we want to drive student learning, then how we assess them is vitally important. What they do on a day-to-day basis will be different according to the demands of the various assessments. A major chunk of the chapter will consider the strengths and weaknesses of the modes of assessment.

I am aware that many academics take what one might call a traditional view, that for lecture-based teaching, the only satisfactory assessment is the timed, unseen, written examination and that students should be left to their own devices on how to drive their own, private work schedule. Indeed, that approach is what many of us experienced as students and the world of academic teaching is nothing if not conservative. Further, there is a consistency to the model and, in most cases, it builds on what students have experienced in school. But the model is built on teaching that is geared towards producing more of the type of people who do the teaching, whereas only a tiny fraction of graduates obtain a post in a university. Most use their graduate skills across a huge range of employment or further study. That skills base makes physicists highly employable but the skills do not develop by accident, which is why there needs to be a diversity of assessment techniques.

Any assessment system incorporates a set of assumptions; these are usually implicit in that many examiners would not recognise them as assumptions because they have become such an intrinsic part of the process. Among them are:

- *Marks are sacred and once earned, cannot be adjusted.* In traditional examinations in number-based disciplines such as physics and mathematics, a student's answer is usually compared with a model answer prepared in advance; to satisfy university rules, those model answers even have a fixed number of marks for each section—we will see some examples in chapter 4. The marks, therefore, merely represent how well the student's attempt matches the template. Average marks often differ for different papers, frequently by large amounts, even when taken by the same set of students. There is nothing unexpected about that; examination setting is not a perfect science. However, there is something wrong if the marks from papers with wildly different averages are combined without adjustment. To be explicit, if paper A has an average of 65% and paper B one of 45%, then simple addition means that, in terms of the combined mark, one is giving far more weight to paper A.
- *On a related point, but within a given module, marks must be added linearly.* I can think of no good reason why this should be the case. Certainly, in the interests of transparency, the way that the module mark is calculated must be made clear but to insist on a simple addition of marks from different components precludes, for example, the incorporation of any competence mastery as a component of assessment. Some examples of how such elements may be incorporated in a fair and consistent manner are discussed in the final chapter.
- *There must be a good spread of marks.* There is sense in making this assumption if the primary purpose of the assessment system is grading. Indeed, some of the grade inflation in degree awards has occurred because the spread of marks has become smaller. The danger in requiring a large spread is that some types of assessment naturally lead to a narrower range. For example, when students come together to work in a group, a typical outcome is that the weaker students learn from the stronger ones. More cynically, perhaps those that have not been doing much work are forced to increase their efforts. Either way, students who are obtaining low marks in traditional examinations usually do much better in a peer-driven environment.

2.2 Norm and criterion referencing

As described in chapter 1, the comparability of degree outcomes between different institutions has been a controversial topic within the UK educational system. Or, to put the debate in the form of a question: are standards the same in all universities? But then we have to ask what we mean by standards (there is often confusion between quality and standards, for some background, see [1]). This question is key to any system of assessment and is concerned with the philosophical approach we take to awarding grades. There are two basic approaches: *norm referenced* and *criterion referenced* and they differ in that the latter claims to offer marks awarded to an absolute standard of achievement, whereas the former awards grades according to the relative achievement of the candidates. I will argue that the main reason why the standards discussion is so confused is that most institutions claim to be awarding marks, and hence degrees, on a criterion-based model but, in practice, are implicitly using norm referencing. As discussed in chapter 6, a recognition of this point could lead to a more rational mechanism of quality assurance across the sector.

2.2.1 Norm referencing

To begin with norm referencing: in this approach, the marks or grades given to a student depend on their achievement relative to the other members of the set of people under assessment. The most common example of a norm-referenced grade is the measure of the intelligence quotient, or IQ. Leaving aside the controversial issue of what IQ tests actually measure, the methodology is to use a series of standard tests and to plot the marks as shown in figure 2.1.

Because the number of people tested is large, one obtains a smooth normal distribution. By definition, the centre of the distribution is classed as an IQ of 100 and anyone within a standard deviation of the centre is deemed 'average'. Those who wish to join the elite organisation MENSA need to be in the top 2% of the national distribution, with an IQ above 130 and those above 140 are officially geniuses.

As a fascinating aside, in absolute terms, the average IQ, in the sense of the actual marks obtained in the tests, is increasing rapidly with time [2]. A person with an average IQ of 100 in 1910 would be rated at 70 today, a value that is considered an indicator of learning difficulties. Equivalently, an average person today would have been rated at 130 in 1910, in the top few % of the population. No one is quite sure why this huge increase has occurred. The change is more or less linear with time and is mainly due to improvements in answering questions that test abstract thinking, rather than those concerned with arithmetic or vocabulary, leading psychologists to ascribe the increase to changes in the world around us and the way we live, particularly with respect to technology (for a review see [3]).

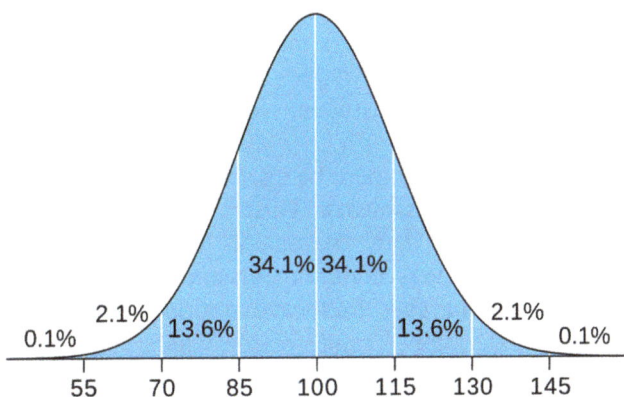

Figure 2.1. Distribution of marks in IQ tests. The vertical white lines represent successive standard deviations. (Published under a CC-BY SA 3.0 licence.)

Norm referencing may be applied at any level within an assessment system, for the final award of the qualification, the mark for a module or course, or even each separate component. To be specific, I will consider the final award and use the traditional UK degree nomenclature: Pass (P); third-class Honours (3rd); lower second-class Honours (2(ii)); upper second-class Honours (2(i)); and first-class Honours (1st). The most common way to apply norm referencing is to define, say, the top 10% of the class as 1st, the next 30% as 2(i), then another 30% as 2(ii), 20% as 3rd and 10% as P. The actual percentages can be changed, of course, so long as they are transparent.

Norm referencing is more subtle at the level of marks for a module. In order for there to be comparability between modules, each set of marks needs to be scaled to a common distribution, with a defined average mark and a standard deviation. In addition, at both the award and modular levels, failures need to be considered on an individual basis. It would be quite wrong to define, say, 5% of the class as failures, regardless of their level of performance. One would therefore need some level of threshold performance before a candidate is included in the scaling.

I am sure that anyone who has worked in the higher education sector bridles at the notion of using a norm-referenced system such as the one described above. However, there are some positive features: grade inflation is removed completely; differences in average raw marks between modules have no effect; and the system operates with transparency and objectivity. The less positive features are listed below:

1. The objection that immediately springs to mind is that the marks are not linked to the learning outcomes of the module. In other words, it is not enough to demonstrate the learning outcomes to achieve a high grade, you also have to do better than your contemporaries. If everyone does super well, there will still be some percentage of people getting third-class degrees. This is a fair point which I will return to later but, in practice, there is always a distribution of marks.

2. Students commenting on such a system would immediately claim that it was not 'fair'. In my experience, they are not usually unhappy if marks are scaled upwards but they are if marks are scaled down. In one institution I am aware of, a student threatened legal action for a mark scaled down. This argument is essentially the assumption that marks are sacred (except when they are low). While that point of view does not stand up to any sort of critical analysis, the power of the student voice means that it does present a significant barrier.

3. Norm referencing relies on there being a well-defined cohort of candidates, all taking the same assessments. This is not usually the case in higher education: some students defer assessments for health or other reasons. Even without deferrals, it is very common for students to take different options and it will not be the case that each option will be taken by students representative of the ability range of the whole class. For example, very mathematical modules often attract high-flying students. This objection strikes me as the strongest argument against a fully norm-referenced system although, as we shall see, it is a difficult issue to resolve within any system.

4. In a benchmarking or competence-based assessment, norm referencing is not appropriate at all. However, I suggest in chapter 6 that such assessments can play an important part in physics programmes.

2.2.2 Criterion referencing

Whereas the principle of norm referencing is that students are graded by their achievement relative to that of their peers, in criterion-referenced grading, candidates' performances are measured against a set of grade descriptors, which in the UK system define what is meant by a first-class grade etc. This is the model that most universities claim to operate and it is possible to find examples of grade descriptors on many of their websites.

One can see the appeal of the criterion-referenced approach. It is easy to link the criteria to the learning outcomes of the module. In addition, a student's individual performance is considered in absolute terms—a student is not affected by how well their colleagues do in the same assessment and, in principle, the whole class could be awarded first-class marks or, indeed, fail.

It is interesting to compare how these criteria are applied within the assessment system in different disciplines. In essay-based subjects, such as English or history, the actual tasks, e.g. essay titles or exam questions, are typically brief and the grading criteria are applied during the assessment, according to the quality and depth of the students' answers. In mathematics and most science subjects, for traditional examinations, there is a much greater emphasis on setting the questions and providing detailed model answers. In these subjects, the assessment criteria are embedded in the model answers, which define the level and standard expected. Consequently, marking tends to be the relatively straightforward task of comparing the students' answers with the model and awarding numerical marks according to the mark breakdown on the paper. Then, at the end of the paper, the marks are added up to a total. There is no explicit comparison with the grade criteria during the marking process.

So far, so good but before we look at some examples of criteria, let us return briefly to the discussion in chapter 1 concerning grade inflation and the comparability of degrees from different institutions. As we saw above, grade inflation is not possible in a norm-referenced system, unless one consciously decides to change the percentage of students awarded, say, first-class degrees. However, comparability between institutions is lost. *In principle*, in a criterion-referenced system, there could be comparisons between institutions—after all, the criteria for a first-class performance are similar across a broad range of institutions. However, in practice, no one believes in this comparability. Further, strong grade inflation has occurred over the last couple of decades, despite the criteria remaining similar over that time. It is possible to argue that students have been working harder or that teaching has improved considerably over that period but I think a more plausible explanation is that the objectivity of the criterion-based approach is an illusion. So, what is wrong?

2.2.3 Examples of grade descriptors

Consider the following three examples of grade descriptors. In each case, the criteria refer to a performance associated with the top grade.

Example A
Understanding
Extensive understanding of key facts demonstrating an ability to formulate ideas in analysis, comprehensive understanding of methodologies with a high degree of precision, highly independent and critical judgement.
Depth of knowledge
Extensive range of sources used and applied to the assignment in a highly insightful manner and of outstanding quality.
Structure
Excellently structured, focused and well written presentation. Compelling argument throughout.

Example B
Candidates will be able to:
- Demonstrate relevant and comprehensive knowledge and understanding and apply these correctly to both familiar and unfamiliar contexts using accurate scientific terminology.
- Use a range of mathematical skills to perform complex scientific calculations.
- Critically analyse qualitative and quantitative data to draw logical, well-evidenced conclusions.
- Critically evaluate and refine methodologies, and judge the validity of scientific conclusions.

Example C
- Displays comprehensive subject knowledge and a thorough command of concepts and principles.
- Selects and applies relevant information, concepts and principles in a wide variety of contexts.
- Analyses and evaluates quantitative and qualitative data thoroughly.
- Constructs detailed explanations of complex phenomena and makes appropriate predictions.
- Evidences great proficiency in solving problems, including those that are challenging or unfamiliar.
- Communicates logically and concisely using appropriate terminology and conventions.
- Shows insight or originality.
- Approaches investigations in an ethical manner, paying full attention to environmental impact and safety where applicable.
- Investigations demonstrate insight and independence to design and complete innovative practical work with highly competent investigative and analytical techniques, and with innovative and effective conclusions to resolve authentic problems.

Example A is designed more generally to cover a range of subjects, whereas B and C relate only to scientific subjects and are therefore more specific. Beyond that, the three examples are roughly comparable. Most academics would concur that to 'use a range of mathematical skills to perform complex scientific calculations', to display a 'comprehensive understanding of methodologies with a high degree of precision' and 'evidences great proficiency in solving problems, including those that are challenging or unfamiliar' are all consistent with what we would expect from a first-class student. Example C, in particular, fits the bill nicely.

And here we see the issue. Example A is drawn from the generic grade descriptors of a leading UK university, describing what is expected of a truly outstanding student. As with all examples from specific universities, I leave the source anonymous. Example B is taken from a description of the top grade of Physics GCSE, an examination typically taken by 16-year-olds in England and which has no mathematical requirements beyond simple arithmetic [4]. Example C describes the requirements for the highest grade of the International Baccalaureate Diploma, an examination taken by 18-year-olds, usually as a qualification for university entrance [5]. Although the examples are extracted from longer documents, I have edited them only in terms of format and not content.

The point here is not that the descriptors are inadequate or wrong, it is that they make no sense without some indication of the level of material to which they refer and without precise definition of what we actually mean by 'insight and originality', 'complex mathematical problems', 'comprehensive understanding of methodologies' or any other of the qualitative virtues mentioned. As a specific example, electromagnetism is a topic taught at all three levels. But, whereas at GSCE (example B), the mathematical manipulations would be restricted mainly to substituting numbers into given equations [6], at university level we would certainly be expecting a confident manipulation of Maxwell's equations using vector calculus.

These examples refer only to examples of criteria for the top grades; below that there are degrees of judgement to be applied. For case B, the IB Diploma, we have, successively, for the top four grades referring to the knowledge and understanding component:

Grade 7: Displays comprehensive subject knowledge and a thorough command of concepts and principles.

Grade 6: Displays very broad subject knowledge and a thorough understanding of concepts and principles.

Grade 5: Displays broad subject knowledge and shows sound understanding of most concepts and principles

Grade 4: Displays reasonable subject knowledge (though possibly with some gaps) and shows adequate understanding of most basic concepts and principles

The reader is invited to muse on the subtleties of the distinction between 'comprehensive', 'very broad' and 'broad' or between 'thorough command' and 'thorough understanding'.

What I hope that I have illustrated is that, while grade criteria are superficially attractive, unless they are presented to an absurd level of detail, any application of them requires both an agreed level of content and, most important, a *subjective* judgement of how the qualitative (and vague) descriptors (comprehensive, broad, thorough etc) are applied. In practice, where this judgement comes from is a comparison of the performance of a particular student with her or his peers, both in the current cohort and in previous years. In other words, what actually determines the grades in this nominally *criterion-referenced* approach, is in fact, a *norm-referenced* comparison with the type of students who are found in that programme in that university. This observation explains why universities across the sector, despite having huge differences in entrance qualifications, turn out roughly similar profiles of degree awards: in short, they grade the students they have against the average level of the cohort.

I mentioned above how, in subjects such as physics, the level of the paper is determined at least as much by the questions that are set as by the way they are marked. Once again, in principle, these questions are set to match the learning outcomes and define an absolute standard, consistent with the criterion-referenced approach. However, in practice, experienced academics will set questions that they know the students will be able to answer to a reasonable level; that is, they use their historical knowledge of the academic capability of their students to set the standard. Consequently, they refer to the norm rather than any absolute criteria.

I have simplified the situation in the interests of clarity. I would also point out that there is much to be said for using historical standards as the basis for assessing the current cohort and I do believe that some sort of approach based on criterion referencing can play a role in an assessment regime more rational than the one we have now. Those ideas will be explored further in chapter 6.

As a final point when discussing grade criteria, if one does apply them, one should avoid the trap of defining excellence and then describing lower grades in terms of departures from that virtuous peak. As a specific example, consider example C above. The definitions provided are for grade 7, the top level. For grade 1, they offer:

- Fragmentary subject knowledge and shows very little understanding of any concepts or principles.
- Rarely demonstrates personal skills, perseverance or responsibility in investigative activities.
- Rarely approaches investigations in an ethical manner, or shows an awareness of environmental impact and safety.
- Investigations demonstrate an ability to undertake very basic practical work with complete dependence on supervised instruction, with attempts at conclusions are either absent or completely incorrect/irrelevant.

This is a dismal list of failure, highlighting what the student cannot do and not what they can. Even at the lowest levels, criteria should require positive features of what has been achieved, however modest those achievements are. Starting the descriptors at the lowest pass levels and describing achievement instead of failure can overcome this problem. As a contrast, consider the GCSE case (example B) where the lowest grade has the following criteria.

Candidates will be able to:
- Demonstrate some relevant scientific knowledge and understanding using limited scientific terminology.
- Perform basic calculations.
- Draw simple conclusions from qualitative or quantitative data.
- Make basic comments relating to experimental methods.

Admittedly, these are limited achievements but they are, nonetheless, achievements, not a lack of them.

2.3 Difficulty and scaling

A former colleague, an Oxford graduate, once told me how surprised he was, on joining the physics department of another university, to find how much more difficult the examinations appeared to be relative to the ones he had taken as an undergraduate. The university, although prestigious, did not have a student intake at quite the academic level demanded by Oxford, and yet those supposedly weaker students were sitting apparently more difficult exam papers. He resolved the paradox when he realised that, at his new university, the same person who set and marked the examination was also responsible for teaching the course; because the lecturer knew precisely what the students had been told, it was possible to set highly complex and detailed questions which relied essentially on rote learning. At Oxford, the papers were set centrally and those teaching the students did not know what the questions would be; the questions were more general, requiring deep understanding but with less specific detail.

The moral of this tale is that it is very difficult to know how challenging an examination paper is by simple inspection. Rather, it is necessary to know what and how the students have been taught and also how they have been prepared for the assessment. The degree of question choice can also play a role—a challenging question has little consequence if no one attempts it.

It follows that setting an examination is not an exact science, although I have known many colleagues who would dispute that statement. If their paper produces a low average mark, they talk about how the class was weaker than usual, despite the same cohort doing perfectly well in other modules. And if the average mark is high, it is ascribed to the quality of the teaching. Such arguments are not convincing.

Given that average marks do vary a great deal between different examinations, there is the question about whether anything should be done about that. In the criterion-referenced approach, in which marks are awarded according to absolute standards, it follows that no adjustment should be made. And in the norm-referenced approach, since the marks are awarded for relative achievement to begin with, there is no need for further adjustment. We are left with the conclusion that in both approaches, we should leave the marks as they are.

This is absurd. The variation in average marks across different examinations for the same cohort (or even from year to year) will be due to a number of factors associated mostly with the impossibility of setting papers with a consistent level of challenge. What the question boils down to is: what do we assume is the constant factor? Do we think that all papers are set and marked with an absolute standard and consistency against the defined criteria and therefore any variation of marks is due to the students' collective lack of ability or effort? Or do we think that the ability and work ethic of the cohort is essentially the same across all modules and from year to year and the variation is due to the setting and marking of the papers? The latter is surely the more plausible.

Does it matter if the average marks differ for different assessments? After all, if the same cohort is taking all the assessments, will there not be an automatic compensation? To some extent there will but consider the following scenario: two papers have average marks of 50% and 70% respectively. Student X gets 90% of the average, 45%, on the first paper and 110% of the average, 77%, on the second paper. Student Y is the other way around, receiving marks of 55% and 63%, respectively. Across the two papers, Student X gets a mean mark of 61% and Student Y, 59%. Without scaling, greater weight is given to papers with high average marks; it is better to do well on an easy paper than on one that is more challenging.

Scaling is not without problems. First, students are rarely bothered about their marks being increased by scaling but rather less enamoured of a reduction. I have known of legal challenges by students whose marks have been adjusted downwards; universities do not like to risk legal examination of their processes.

There are also technical objections. Different assessments lead to different mark ranges; for example, a highly mathematical, unseen examination will have a much broader range of marks than will a continually assessed laboratory module, a difference that can be justified by the nature of the assessment. More seriously, scaling only works consistently if the same cohort takes all the assessments. This rarely happens: students take different options and students often take assessments at different times, either because they are resitting or missed the original assessments due to ill health. In the former case, it is by no means obvious that each option should produce the same average mark; for example, more mathematical options often appeal to the more able students. In the latter case, the students are actually taking a different assessment from the main cohort and, also, it would not be correct to undertake a separate cohort analysis both because the numbers are small and because the students do not form a representative group.

These are thorny issues. Some suggestions on how to deal with them are explored in chapter 6.

2.4 Types of assessment

The discussion to date has mainly comprised general considerations about the purposes of assessment and the philosophies that underlie it. In this section, I present different methods of assessment, with a focus on the types that are most relevant in a physics context, although most of them are more widely applicable.

The methods are classified into four groups:
- Formal, timed assessments.
- Continual assessment.
- Real-time assessment.
- Other.

A strengths and weaknesses analysis of each group is presented in tables 2.1, 2.2, 2.3 and 2.4. In each case, the method is linked to the purposes in chapter 1 and evaluated in terms of the broader assessment framework. The emphasis in this section is on assessments applied to theory courses; assessments associated with student-centred activity will be explored further in chapter 5.

Table 2.1. Strengths and weaknesses of timed assessments.

Formal timed assessments (discussed further in chapter 4)

Type	Description	Strengths	Weaknesses
1. Traditional unseen examinations	Students usually sit in an exam hall and have no access to their notes or any other sources of information. Usually involves combinations of bookwork and problem solving in relatively long questions ~ 30–40 min.	• Wide distribution of marks so good for grading. • Well understood across the sector. • Every student has the same examination. • Easy to mark consistently. • Efficient in staff time. • Hard to cheat.	• Severe restriction on what can be assessed (extended problem solving, computing, research skills etc). • Does not promote active learning. • Tends to encourage rote learning.
2. Open-book examinations	Students sit an unseen paper as above but have access to their notes and textbooks. The examination is usually in an exam hall, although the Covid-19 crisis did mean that assessments were taken at home for many students.	• Wide distribution of marks so good for grading. • Every student has the same examination. • Easy to mark consistently. • Efficient in staff time. • Less dependence on rote learning.	• Severe restriction on what can be assessed (extended problem solving, computing, research skills etc). • Does not promote active learning. • If not in examination conditions, easy to cheat.

(*Continued*)

Table 2.1. (*Continued*)

Formal timed assessments (discussed further in chapter 4)

Type	Description	Strengths	Weaknesses
3. Multiple choice questions	Questions where students have a choice of four (usually) answers, one of which they select. Test can be taken in class, in an examination hall or online.	• Wide distribution of marks so good for grading. • Covers a wide range of material. • Good for testing conceptual understanding of key concepts. • Marking is automatic. • Good for instant feedback.	• Relatively superficial questions. • Need to address issue of random guesses. • Hard to set unambiguous questions. • No credit given for partial working.
4. Short questions	A variation on the traditional, timed examination in which there is a set of questions, each requiring a short answer ~ 10 min. Such questions sometimes form a section within a timed examination.	• Wide distribution of marks so good for grading. • Can cover a wide range of material. • Easy to mark.	• Can only ask relatively superficial questions. • If a section in a traditional unseen examination, students often spend too much time on these questions.

Table 2.2. Strengths and weaknesses of continual assessment.

Continual assessment

Type	Description	Strengths	Weaknesses
1. Essays, dissertations and literature reviews	Students are given a topic and produce an essay or review. Assessed elements may include oral presentations, plans etc, as well as the final document	• Allows students to study a topic in depth. • Develops important transferable skills. • Can be linked to experimental and other work (see chapter 5).	• In physics, it is hard for an undergraduate to say anything original about research physics, so essays tend to be syntheses of other work…. • …hence a thin line between originality and plagiarism. • Hard to mark consistently.
2. Reports and research papers (discussed further in chapter 5)	Reports of some sort of student-led activity such as a project or other investigation. May take a number of forms, including a formal research paper, and may be linked to other assessments such as oral presentations.	• Students able to demonstrate originality. • Develops important transferable skills. • Easy to embed feedback.	• Marking is labour intensive. • Difficult to achieve consistency of marking.
3. Problem sheets	Regular problem sheets which can be distributed informally as formative exercises, often alongside optional workshops, or with formal deadlines and a summative contribution.	• Encourage students to keep on top of their work. • Provide feedback on progress to both student and teacher. • Students talk to each other about physics.	• Unless the marks count, student participation can be very low. • Hard to eliminate cheating by students copying answers. • With, say, weekly problem sheets, there is a heavy marking load.
4. Group work (discussed further in chapter 5)	Students working in groups to accomplish tasks. These can range from mini-projects associated with specific modules to large projects or investigations. The assessment format may include any of the three above types.	• Involves teamwork and the associated skills, which are highly valued by employers. • Encourages students to talk about physics. • Ensures students participate. • Can experiment with peer assessment.	• Hard to assess individual contributions. • Difficult to achieve consistency of marking.

Table 2.3. Strengths and weaknesses of real-time assessments.

Real-time assessments (discussed further in chapter 5)

Type	Description	Strengths	Weaknesses
1. Oral examinations (vivas)	Students are interviewed individually. The content may be related to a specific module, a project, or physics in general. An essential part of a PhD examination.	• Specific to the student so can be useful for group work. • Very flexible: allows most able students to go into greater depth, for example. • No possibility of cheating.	• Students find 1-2-1 interviews very stressful. • Very labour intensive. • Limited coverage of material. • Consistency very difficult to achieve.
2. Oral presentations	Oral presentations are usually live, with an audience and are usually associated with some work that the students have done, for example a project. Typically, students answer a few questions at the end.	• Important transferable skills. • Ensures student participation. • Can be recorded to aid marking. • Can include peer assessment.	• Some students become highly stressed at speaking in public. • Very labour intensive. • Marking consistency very difficult to achieve.
3. Podcasts etc	Students create online presentations, which can incorporate a talking head, animations, video clips etc.	• Encourages creativity. • High levels of student enjoyment. • Development of skills for employability. • Podcasts can be used for other purposes (e.g. student recruitment).	• Marking consistency very difficult to achieve, particularly with different formats. • Labour intensive.
4. Posters	Either alone or in a group, students create a poster which is displayed at a dedicated session alongside similar efforts from others.	• Encourages creativity. • High levels of student enjoyment. • Development of skills for employability. • Posters can be used for other purposes (e.g. student recruitment). • Can be marked by one person.	• If there is a group, difficult to separate contributions. • Marking criteria may lack specificity with a broad range of content.

Table 2.4. Strengths and weaknesses of other forms of assessment.

Other methods of assessment

Type	Description	Strengths	Weaknesses
1. Computer program	A program, or section thereof, forms a natural assessment for a course on computing but numerical solutions or modelling can also be used within physics modules.	• Promotes computing skills. • Enables a broader range of problems to be set than those just with analytical solutions. • Can be used as a benchmark assessment.	• Not always easy to provide grades. • Authenticity issues—easy for students to copy. • Harder to mark than simple written work.
2. Competency progression	Students work their way through material and are allowed only to progress once they have demonstrated competency. New possibilities with this approach with AI and contingent pathways.	• Assessment intrinsic part of teaching and learning and matched to the individual student. • Can be interactive. • Students are required to understand material as they go along. • Links well with benchmarking.	• Students progress at different rates and not all students will complete assessments. • Still experimental in terms of its online potential. • Does not easily lend itself to grading.
3. Synoptic	An assessment, often a formal examination, which is not linked to specific content.	• Test of physics as a synoptic whole. • Rote learning plays no part.	• May be difficult to accommodate in a modular system. • Although the principle is authentic, the format of a written examination is not.
4. Portfolios and artefacts	Students provide a selection of items in a loose portfolio framework. Could be anything from a working electronic circuit to a selection of material developed as part of an industrial or other external project.	• Encourages students to be imaginative. • Allows items to be included that do not fit with other types of assessment. • Fits with benchmarking.	• Difficult to mark consistently. • Possible issues with authenticity.

2.5 Students with disabilities

The *Equality Act 2010* [7] requires universities to make 'reasonable adjustments' to accommodate disabled students. That accommodation includes assessment and the various methods outlined above provide a rich source of possible alternatives, according to the specific disability.

It is worth mentioning that students entering university have already spent 13 years or so in the education system. They need sympathetic support but, in most cases, they know how to cope. It is also worth bearing in mind that graduation carries a higher premium for disabled people: relative to those with only GCSEs, disabled graduates are 24% more likely to be in employment (73% versus 49%), compared with 13% for those without a disability [8].

The phrase 'reasonable adjustments' immediately raises the question of what constitutes 'reasonable'. Referring back to the cartoon in figure 1.1, one might have to accept that the laudable ambition of the goldfish to climb the tree might not be capable of being accommodated. But these exceptions are few and far between. Universities are *legally* required to make reasonable adjustments; it follows that if, say, one university can accommodate a student with a particular disability, it would be hard for another institution to make a case that no reasonable adjustment could be made in a similar case.

The definition of disability in the 2010 act requires a physical or mental impairment that has a 'substantial' and 'long-term' negative effect on the person's normal daily activities. 'Long-term' is usually interpreted as being for at least a year. The definition of 'substantial' is less clear-cut but it normally applies when someone has difficulty in completing everyday tasks.

In recent years, mainly for the very positive reason that there is less stigma attached to mental illness, the numbers of student entering universities with a reported disability has increased markedly. The Office for Students in England reports [9]:

> Between 2010 and 2017, the proportion of students in England self-reporting a disability increased from 8.1 per cent to 13.2 per cent. Much of this increase has been driven by students reporting mental health issues, the incidence of which has grown from 0.6 per cent of all students in 2010 to 3.2 per cent seven years later. The proportion of those reporting a social or communication impairment has more than trebled from 0.1 per cent to 0.5 per cent of all undergraduate students over the same period.

Perhaps partly as a result of this increase, most universities now have robust structures in place for supporting students with disabilities: some of these may operate centrally, such as the provision of laptops or braille translations of notes and textbooks, but most of the adjustments necessarily happen at the subject level.

Considering assessment explicitly, there are three principles that I recommend in this context:
1. Adjustments should be made to assist disabled students to achieve the required learning outcomes; other than in exceptional circumstances, the learning outcomes themselves should not be altered.

2. Any adjustments to teaching or assessment should be inclusive. The experience of disabled students should be as close as possible to that of the other students.
3. The students themselves are usually the experts in coping with their disability and should be involved in any discussions about assessment. It is essential that any special arrangements for assessment are formally agreed well in advance.

There is a wide range of disabilities that students might have and each presents a different barrier. The list below is not exhaustive but covers many of the more commonly encountered disabilities. In each case, I offer some comments on how adjustments might be made, bearing in mind the principles above. I consider only assessment, not more general adjustments (a general guide to helping students with disabilities is [10]). In addition, bear in mind that no two students are alike.

2.5.1 Specific barriers to learning (e.g. dyslexia or attention deficit hyperactivity disorder etc)

For timed examination sessions, most of the students in this category can be accommodated without undue difficulty, usually by allowing extra time in a separate examination room.

For continually assessed work, particularly when it requires the submission of an extended written document, the situation is less clear-cut. A natural extension of the policy for examinations might suggest that the student be given an extension to a deadline and that may, indeed, be a sensible move in some cases. However, there is a serious risk that, by postponing deadlines, a student can be placed in a situation where, for example, their preparation time for examinations is compromised. My own preference, based on experience, is that the best approach is to stick to the formal deadlines but lend a sympathetic ear to special cases. The emphasis should be on helping students with disabilities cope with the assessment regime, wherever possible. Meeting deadlines is an important skill.

Where the continually assessed work is something such as the weekly submission of solutions to problems, again, the best arrangement is to keep to the deadlines, while ensuring that the student knows that help is available.

With the government placing an increased emphasis on employability skills, many physics programmes include modules that improve oral and written communication skills and the assessment for the latter will typically include marks for (reasonably) correct grammar. For dyslexic students, that is very challenging. This issue falls into a more general category where the learning outcomes for the course may be especially difficult, or even impossible, to achieve for students with certain disabilities and that point is discussed further below.

2.5.2 Physical disabilities affecting mobility or dexterity

Although these disabilities can have a profound effect on one's quality of life, they are usually less of a difficulty in assessment terms, provided access issues are addressed.

In some cases of reduced dexterity or students with conditions that prevent them sitting still for long periods, there will be a need to increase the time available for a formal examination. Issues around laboratory work for people with reduced dexterity are discussed below.

2.5.3 Sensory impairments such as deafness or severe loss of sight

The number of young people with some sort visual impairment in the UK is relatively low at 0.16% (DfE data can be accessed at [11]) although the RNIB reports that 1 in 2000 has a serious impairment. I am personally aware of several physics graduates who fall into this category. Inevitably, any student that has managed to satisfy the entrance requirements for a physics programme will have already found ways of coping with written work and, to some extent, mathematics and diagrams. It is essential that any support, including the use of software, is based around the strategies already developed and the same consideration applies to assessment. In passing, I note that, for the students I know about personally, the biggest support issues have not occurred in the departments but with the central units failing to deliver, for example, on converting texts to braille or providing suitable examination papers.

Some visually impaired people have enough sight to read suitably large print and current software can also convert the written word into spoken language or braille. However, for any assessment, there are two essential adjustments. First, there must be extra time allocated for formal examinations; a student with good vision can scan a paper quickly, an option not available to visually impaired colleagues. Second, where at all possible, there should be someone on hand to assist the student in understanding the task set. Diagrams and equations, in particular, need to be described so the person assisting must have a knowledge of physics at a higher level than that under assessment. It is good practice for that person to be someone already familiar to the student in this context. This adjustment applies just as much to problem sheets etc as to examinations.

It is worth stating that the overarching principle of inclusion is highly relevant for visually impaired students. They can and should participate in group projects and can perfectly well give oral presentations and make podcasts. Also, because severely visually impaired students are relatively rare, it is a reasonable adjustment that all assessments of a cohort that includes such a student to be set sympathetically to their requirements. That is not to say that they should be set a different paper but that the paper for everyone in the cohort should not, for example, require detailed inspection of a diagram or an equation.

Around 0.05% of young people in the UK are classified as deaf [12] but rather more will have some kind of hearing loss. Modern hearing aid technology allows most of the latter to cope well with everyday interactions and there are no particular issues with written assessments. The type of everyday adjustments that are used in teaching, such as subtitles for lecture capture and speakers in live sessions facing the student in question, can also be applied to assessment environments.

Profoundly deaf students are rare in physics and in higher education generally, although I have met one such graduate. Any student that is admitted to a university will have some means of communicating with the world, either by lip-reading or by using sign language. In the former case, provided we adhere to the principle of involving the student in the discussions, assessment of any sort should not pose major problems beyond those encountered in the teaching.

For students that use sign language, a minimum requirement would be a sign-language interpreter. However, as has been documented (for a description of the challenges faced see [13]), interpretation is far from trivial in such a conceptually rich subject. Sign language requires development of signs for new words and concepts and, while basic ideas such as mass, energy, charge etc, have signs that would be known to any user, the plethora of advanced jargon requires the development of new signs which, to my knowledge, do not currently exist. In addition, the sign interpreter would themselves need a physics background.

2.5.4 Social or communication impairments, such as autism

In recent years, a great deal has been done to improve the university experience for students on the autism spectrum [14]. In terms of assessment, it is common to allow students to take timed examinations in separate accommodation, preferably in a room with which they are already familiar. Rather less common, but nonetheless important, is the use of language in setting assessments of all types. Students with autism require unambiguous instructions. The best approach is to work with autism advisors to ensure that the language used will not lead to problems. In addition, as is the case for all students with disabilities, speaking to the students, making them feel included and part of the decision-making process will overcome many of the issues.

2.5.5 Mental health conditions (e.g. anxiety, phobias, anorexia, depression)

As mentioned above, much of the increase in disclosed disabilities has been in this area, mainly for the positive reason that students feel disclosure carries less stigma than might have been the case in the past. Assessment poses significant anxiety problems for students with these conditions, whether the artificial environment of a timed examinations, which leaves all students anxious, or the need to meet submission deadlines. Unfortunately, there is no easy way to modify assessments to make them easier to cope with, since it is the fact of the assessment rather than its form which leads to the anxiety. As a consequence, it is common for students with these types of disabilities to repeat assessments and even to miss whole years. Good practice can be to reduce the density of assessments where possible, even to the extent of the student working part-time to an agreed plan. Some advisors recommend reducing the number of assessments but I remain to be convinced by that approach as it increases the significance of those that remain. As always, the mantra is that no two students are alike so it is essential to work with the student to find the best way forward.

2.5.6 Long-term health conditions, such as cystic fibrosis

To be considered a disability according to the Act, a health condition must last a year or more. However, any serious health problem, such as a broken leg, will have a disruptive effect on a student's progress. In terms of assessment, that usually means a simple postponement until the student recovers. Longer term issues, such as cancer treatment, usually follow a similar path, albeit on a longer time-scale.

Some students have ongoing chronic conditions which they will have throughout a large part of their life. I have taught many students in this category, including two with cystic fibrosis, a very debilitating condition that usually requires many hours of daily treatment. In neither case were any adjustments made nor any requested, except a separate room for timed examinations, with suitable time extensions. In general, such physical disabilities do not require substantial changes to the assessment regime.

2.5.7 Disability and learning outcomes

The way universities work with disabled students is laudably positive and huge improvements in awareness have been made in recent years. However, there are a couple of issues that are often not discussed, due to their negative connotations. I feel that is better to address them openly. The first relates to learning outcomes, which are, quite correctly, drawn up in terms of both knowledge and skills. The latter includes both subject-related skills and those related to employability, such as teamwork and the ability to communicate to a broad range of audiences. Assessments are necessarily tuned to these outcomes but, in some cases, a student with a disability may not be able to achieve the required outcome.

It is helpful to consider specific examples. I have taught several modules that try to enhance communication skills and in which students are expected to be able to communicate in clear and unambiguous English. Evidently, dyslexic students will struggle with this requirement but, if we return to the list of principles, the solution here is to provide full support to the student but not to adjust the marking criteria.

Other examples can be more difficult to resolve. A blind student, or someone with problems in manipulation, is clearly going to suffer in a laboratory environment and yet, not only is some appreciation of experimental work an essential part of the training of a physicist, in the UK and Ireland it is also a necessary condition for the degree to be accredited. In these cases, it is probably best to deal with the laboratory component by arranging for the student to have a helper, who carries out the manipulation following the instructions of the student. There are also an increasing number of virtual laboratories that provide a quasi-realistic experience. While it is probably the case that students with these types of disability are better suited to the more theoretical side of physics, it is perfectly straightforward to put together and assess a module based on practical work and there are many examples where this has happened [15].

A third example lies with students who have a mental health condition that, for example, makes it difficult for them to interact with other people in a group project, or to use apparatus that other people have touched, or to stand up to give an oral presentation. As with all cases involving disabled students, the first thing to do is to talk to the student to see what compromises can be achieved: for example, in some cases, students are happy to give a presentation to a couple of staff but not to a group of their peers. Or they might be comfortable working in a group of students they know well rather than relative strangers. Nonetheless, there will be occasions when the disability is such that some learning outcomes cannot be achieved and there is no natural substitute. One is then faced with the dilemma of either failing the student or compromising the assessment criteria. Sensibly and pragmatically, the latter course of action is almost always taken; it is patently absurd to say that someone is not worthy of a physics degree because they are not able to speak in front of large audiences. How that is achieved can be far from trivial, however. In the words of a document providing guidance to assessment at school level [16]: 'At present it is not always clear where the line should be drawn between adapting assessments and providing alternative assessments for pupils with severe special needs.'

Such considerations bring me to the second issue which is seldom discussed, which is to what extent should special arrangements for disabled students be extended to all. Where the arrangements are simply allowing extra time and/or being housed in a separate room, the answer is: not at all. The whole point of those concessions is that disabled students are given more time *relative* to the other students in order to level the playing field. But where a specific assessment—perhaps an oral presentation—might be substituted by a piece of written work for a student with a disability, should that substitution also be available to all students? My own view is that it should not be, but it is an arguable point.

Another way in which arrangements used to support disabled students has impinged upon assessment is the common requirement that a complete set of lecture notes, plus answers to any problem sheets, be available to students, sometimes in advance of lectures being given. Such an arrangement has created the notion of a canon of material that needs to be (rote) learned rather than lectures being a springboard to further, student-driven learning: it is a common complaint in student surveys, including those that are visible external to the university, that the assessment did not follow the lecture notes. Such arrangements can also reduce class attendances, particularly when lecturers stick rigidly to the notes. Here is not the place to discuss how a more interactive teaching environment can change this situation but it is clear that an imaginative approach to assessment can have a profound effect in shifting away from the idea that the lecture notes are a full definition of what has to be learned.

To finish this section, I briefly return to the principles identified at the beginning. The relationship between teachers and disabled students should be as open as possible and the student should be involved in any plans for changes to the normal diet. In parallel, the assessment environment should be as inclusive as possible, with the emphasis on helping students cope with the existing assessments by offering them support.

2.6 Concluding remarks

This chapter has considered the types of assessment that are used in physics programmes, considering their strengths and weaknesses relative to the requirements of any system of assessment, as described in the final section of chapter 1. It also discussed the two principal philosophies underpinning the process by which degree outcomes are determined. Lastly, there were also some remarks on how students with disabilities can be accommodated fairly and sensitively within the assessment regime.

Three broad tensions have emerged:
- *Promoting learning versus authenticity.*

  The types of assessment that best promote student learning often involve group work, projects and other continually assessed tasks; these can be difficult to mark consistently and there might be problems in distinguishing the contributions of individual students. Timed examinations, in contrast, although very limited in the type of questions they can ask, and encouraging rote learning, are easy to mark (saving staff time—another important consideration), lead to consistent outcomes and provide a good spread of marks.
- *Norm referencing versus criterion referencing.*

  Here we have the curious situation that, while universities claim to operate a criterion-referenced system, in practice they apply norm referencing to determine absolute grades. In other words, while the defined criteria are certainly useful in distinguishing the relative performance of students within the cohort, the translation into absolute marks or degree classifications is norm referenced in the sense that examiners base their judgements on the type of students they are used to.
- *Student view versus educational goals.*

  Put simply, students seek the best qualification, not necessarily the best education, although the two, of course, are not incompatible. To exaggerate the point, they have a vested interest in the most straightforward types of assessment and are likely to comment adversely on anything that they find challenging, even though they might have learned more physics through that route. In the UK, this tension is exacerbated by the student views being in the public domain, via the NSS and TEF, and provides strong encouragement for grade inflation and for discouraging innovative assessment. These tensions are touched upon in chapter 4, which looks at examples of assessments and will also be discussed in more detail in chapter 6.

Bibliography

[1] UK Standing Committee for Quality Assessment 2017 Protecting the comparability of degree standards *UK SCQA* https://ukscqa.org.uk/what-we-do/degree-standards/
[2] Flynn J 1994 IQ gains over time *The Encyclopedia of Human Intelligence* ed R J Sternberg (New York: Macmillan) pp 617–23

[3] Trahan L, Stuebing K K, Hiscock M K and Fletcher J M 2014 The Flynn effect: a meta-analysis *Psychol Bull.* **140** 1332–60
[4] Ofqual 2017 Single sciences (biology, chemistry and physics) and combined science: grade descriptors for GCSEs graded 9 to 1 *Ofqual* www.gov.uk/government/publications/grade-descriptors-for-gcses-graded-9-to-1/grade-descriptors-for-gcses-graded-9-to-1-single-science-biology-chemistry-and-physics-and-combined-science
[5] International Baccalaureate Organisation 2017 Diploma Programme: Grade Descriptors *International Baccalaureate Organisation* www.ibo.org/contentassets/0b0b7a097ca2498ea50a9e41-d9e1d1cf/dp-grade-descriptors-en.pdf
[6] Wong V 2018 The relationship between school science and mathematics education *PhD Thesis* King's College London https://kclpure.kcl.ac.uk/portal/files/95881991/2018_Wong_Victoria_1231276_ethesis.pdf
[7] UK Government 2010 *The Equality Act*
[8] UK Parliament 2021 Disabled people in employment *House of Commons Briefing Paper* Number 7540
[9] Office for Students 2019 Beyond the bare minimum: are universities and colleges doing enough for disabled students? *Insight Brief* www.officeforstudents.org.uk/publications/beyond-the-bare-minimum-are-universities-and-colleges-doing-enough-for-disabled-students/
[10] IOP 2008 Access for all: a guide to disability good practice for university physics departments *Institute of Physics Guide* www.iop.org/sites/default/files/2021-03/access-for-all-disability-good-practice-university-2008.pdf
[11] VIEW 2021 DfE official data—children and young people with VI—England *VIEW*
[12] Consortium for Research in Deaf Education 2019 UK-wide summary *National Deaf Children's Association* www.ndcs.org.uk/media/6550/cride-2019-uk-wide-report-final.pdf
[13] Hendrickson J 2019 On becoming a physicist: Colin Lualdi shares the challenges and triumphs of a deaf physics graduate student *Physics Department, The Grainger College of Engineering, University of Illinois Urbana-Champaign* https://physics.illinois.edu/news/article/34827
[14] Hughes M, Milne V, McCall A and Pepper S 2009 Supporting students with Asperger's syndrome *A Physical Sciences Practice Guide* Higher Education Academy Physical Sciences Centre www.studynet2.herts.ac.uk/ltic.nsf/Teaching+Documents/5C45A9966D537C2780257E96004E9AF1/$FILE/supporting_students_aspergers_syndrome_rpg%5B1%5D.pdf
[15] Dunn J and Morgan L 2009 Teaching a physics laboratory module to blind students *Physical Sciences Centre Tool Kit* Higher Education Academy https://heacademy.ac.uk/system/files/teaching_physics_laboratory_module_2009.pdf
[16] Cumming J J 2008 Legal and educational perspectives of equity in assessment *Assess. Educ.* **15** 123–35 cited by Wilkinson K and Twist L 2010 *Autism and Educational Assessment: UK Policy and Practice* (Slough: NFER)

IOP Publishing

Assessment in University Physics Education

Peter Main

Chapter 3

What are we assessing?

3.1 Introduction

In the first chapter I introduced the purposes of assessmentze. In brief these were:
- Classifying students or grading.
- External reputation and quality assurance.
- Promoting and measuring learning.
- Benchmarking.

In chapter 2 I introduced some of the methods of assessment and looked at their strengths and weaknesses with regard both to the purposes and to the principles underlying them, which are:
- Validity.
- Appropriateness and transparency.
- Authenticity.
- Accuracy and consistency.
- Achievability.

Throughout the discussion I have made frequent reference to 'learning outcomes' without either explaining what I mean by the idea or producing any detailed examples. In principle, learning outcomes are more pertinent for curriculum design than for assessment. When designing a new programme or a new module, the starting point is to decide what students should be able to do at the end of it. This sounds very straightforward, obvious even, but many teachers struggle with the task, often for the (laudable) reason that they see their role simply as helping students understand physics. They do not start from questions such as: what do we want physics graduates to be able to do, or how does my module contribute to the overall programme in that respect?

Notice that I place an emphasis on what students can do—the learning outcomes—and that implies being measurable or, if you prefer, capable of being assessed.

For that reason, it is best to avoid terms such as 'appreciate' or 'understand', which are vague and subjective. As a concrete example, consider a lecture-based module on electromagnetism, which contains some material on Gauss's theorem. A suitable learning outcome might be: 'to apply Gauss's theorem to determine the electric field and potential of charge distributions of high symmetry'. Or one might make reference to using numerical techniques to solve more complicated charge distributions, and so on.

The learning outcomes should be designed into the programme, along with the content. The link to assessment is then straightforward: whatever the students are expected to be able to do should be tested in some way, although not every single topic has to be tested every year. In the next section, I concentrate on the learning outcomes that might be appropriate at the programme level and what types of assessment are appropriate for each of them. Note that programme-level outcomes do not have to occur in all modules or courses; they can, and will, be distributed across several. For example, oral presentations can be included as part of the assessment for many courses but I would argue that the assessments should be coordinated for consistency, for example in terms of their marking criteria, to ensure the students receive a common message. In my experience, this coordination rarely happens.

Before moving on I wish to comment briefly on a couple of the assessment purposes listed above. First, the use of assessment or, more accurately the subsequent grades, for quality assurance purposes is not relevant for a discussion on student capabilities and learning outcomes; I will return to that purpose in chapter 6. A more relevant purpose here is the notion of benchmarking, a type of pass/fail assessment which is readily applicable to physics and other sciences but which is rarely used in university physics departments. It might apply when there is a specific task or exercise that must be completed in order for the outcome to be achieved but where there is little sense in degrees of performance. An example might be, say, the ability to use a piece of equipment or to apply a particular algorithm using a designated computer language. My view is that benchmark assessments can play a role in physics programmes and this suggestion will be discussed in more detail in chapter 6.

3.2 Testing physics programmes

A degree in physics is a very marketable qualification. Since everything is subject to physical laws, it is common to find physicists working in many different areas of science and engineering. One often hears people say that there is no physics industry, compared with, say, chemistry, but I counter that by saying that, when physics becomes useful, we call it something else. The vast majority of diagnostic techniques in medicine are physics-based, including magnetic resonance imaging, positron–electron tomography and brain scanners based on superconducting devices, as well as more basic probes such as x-rays, ultra-sound and radioactive tracers. The global positioning system (GPS) is not a theme in any physics department in the UK but it involves the use of atomic clocks, suitably corrected for relativistic effects.

However, physicists are not only in demand due to their knowledge base; it is also their way of thinking, their flexibility and their approach to problem solving that makes them distinctive and so valuable across a range of employment, including areas where the degree content is largely irrelevant. Employers like to recruit people who are mathematically and computer literate and who are used to problem solving. And while employers in a management context might not require their staff to use quantum mechanics, the fact that they have grappled with the topic shows an ability to grasp difficult concepts coupled with an admirable perseverance.

The point here is that it is what a physicist can do, at least as much as what they know, that defines the appeal of a physics graduate. A few of their skills might be specific to the subject but most are generic and applicable to many different environments. The role of assessment is not only to measure to what level students have acquired these skills—the grading purpose—but also to drive their acquisition and development.

In what follows, I am going to state what I think are the essential properties of a physics graduate. In each case, I state briefly why I have included the property and then discuss what types of assessment, drawn from those listed in chapter 2, would be helpful to use. I also comment on any issues that can arise and how they might be overcome. I shall not offer suggestions for the physics content; that is for the institutions to decide and different degree programmes have different flavours. My only comment is that there is an understandable tendency to introduce more and more material but without removing an equivalent amount. Very few physics graduates become university academics and it is their skills base that allows the majority to pursue a successful career. Content is important but it needs to be limited to a manageable quantity.

The criteria I have used to decide what to include in the list of learning outcomes are:
- Relevance to a physics programme (e.g. a foreign language is a useful asset but would fail this criterion) *and*
- Skills associated with being a physicist *or*
- Skills that are readily applicable in a variety of circumstances.

3.2.1 Knowledge

Although physics is better defined as a way of thinking rather than in terms of specific content, it is clear that, to operate successfully as a physicist, one does need to have a knowledge base on which to draw. It is outside the remit of this book to say what that should comprise but international comparisons show broad agreement on the most important areas. For those in the UK, the IOP's accreditation documentation [1] and the QAA Subject Statement for physics [2] both provide a sensible consensus.

Assessments
The most effective way to assess a knowledge base is by a timed, unseen examination. The format of that examination can include:

- Multiple choice questions, which are ideally suited to testing very basic knowledge and qualitative reasoning.
- Short answers, where something more than a simple statement is required.
- An element of a longer question, sometimes merely to signal to the student which part of physics is relevant.

An alternative that is popular in some countries is an oral examination. However, in the context of testing knowledge, there appear to be no obvious benefits over a written test and the substantial disadvantages of the increased workload for staff, the difficulty in being consistent and the stress on the students.

Commentary
I commented earlier on the overuse of rote learning in university physics examinations. In devising assessments, examiners should consider very carefully what they expect a student to remember. For example, learning the derivation of a standard formula by rote, then reproducing it in an examination room has limited educational benefit. An interesting point in this context is the balance between expecting students to know material and them being able to access material. Evidently, both are important but the latter is perhaps more useful in the longer term. Indeed, in a timed examination, it may well be the case that an open-book format is just as effective as the unseen version, even where there are straightforward tests of memory, because the students who do not have the material at their fingertips will waste time searching for it.

In some subjects, where a specific knowledge base might be crucial, such as the medical sciences or, say, training for various trades, such as electricians and plumbers, a competency or benchmark assessment is appropriate. In the case of the plumber, an example might be to be able to install a central heating system. Such assessments are not common in physics but there is no reason why they should not be used; if specific knowledge is deemed essential, then a pass/fail element to the assessment is perfectly reasonable. The barrier to that approach is how such a digital outcome is incorporated in the final grade; this issue will be discussed further in chapter 6.

As with all the items on this list, one of the relevant purposes of assessment is to measure learning, for the benefit of the teacher as much as for the student. While summative assessment of the knowledge base is best achieved via some sort of timed examination, in-class MCQs can provide the teacher with instant feedback on how students are progressing.

3.2.2 Mathematical skills

Mathematics is the language of physics and all physics programmes contain training in mathematical techniques. Typically, mathematics enters the curriculum in two ways: first, via explicit training in general mathematical methods, which usually occurs in the early years, and second, within more advanced physics courses, such as, say, general relativity, which require more sophisticated techniques. In addition, projects may have a mathematical flavour.

Assessments
Timed, written examinations are a good method of assessing whether students have the requisite skills to solve straightforward problems. There is no obvious benefit or disadvantage in making the examinations unseen; if a student is not fluent in a mathematical approach, they are not going to be able to learn it within the two or three hours of the session.

Problem sheets are also very useful assessments for mathematics. A common observation made by students arriving at university is the difference in the way mathematics is taught. In their school or college, they are given problem after problem to ensure fluency in a single technique. By contrast, in university, the density of material in the lectures is much higher and there are relatively fewer problems assigned. Given the reluctance of some students to attempt tasks that have no summative element, there is a great deal to be said for regular, marked problem sheets, the primary purpose of which is formative, but which retain a small summative element.

Projects may require specific mathematical skills but given that, by their nature, projects are specific to individuals or groups, they are perhaps not best seen as an assessment of mathematics *per se*. Mini-projects, however, may be a helpful way of encouraging students to learn from each other, although regular problem sheets probably achieve the same aim.

Commentary
Multiple choice questions are not suitable for testing mathematical skills although short-answer questions might be useful at early stages in order to cover more of the curriculum. Oral examinations are similarly inappropriate.

An interesting point that, in my experience, is rarely addressed in physics programmes is whether students should be given guidance on how to learn new mathematics for themselves rather than apply mathematics taught by someone else. Mini-projects that rely on a student or a group of students digging out new ideas can be effective in this context.

If some mathematical skills are deemed essential, it may be that some form of competency or benchmark assessment would be appropriate.

3.2.3 Computing

The ability to program is an essential part of a physicist's training. Although there is a great deal of variety in terms of which languages or high-level software are taught, most graduates are expected to be able to write code of some sort and to be able to use a range of numerical techniques to solve equations and/or develop models and simulations.

Assessment
The only way to assess computing skills is for students to write programs to solve problems. The great beauty of the topic is the spectrum of levels that can be encompassed, from the use of simple algorithms to simulations of complex systems.

Projects and mini-projects for individuals or in teams are simple to invent. Written examinations of almost any sort are unsuitable for testing computing skills.

Commentary
Computing is an area where competency assessments are highly relevant and it is entirely plausible for the material to be taught in such a way that a student moves to a new topic only after having demonstrated mastery of the previous one [3].

A particular issue when assessing computing skills is that, even before any instruction takes place, there is a massive range in students' starting points. Some are already able to write code fluently, whereas others have IT skills no more sophisticated than using a basic spreadsheet. There is no simple but fair way of dealing with that range. One can offer everyone the same teaching and assessment, in which case all those with substantial prior experience will be wasting their time, albeit probably receiving high grades. Alternatively, one could run parallel classes, which allows those with better preparation to go further; in a mastery model, such as that mentioned above, this would happen naturally. In such a case, the better prepared would be using their time more productively but it would mean that, in effect, the different groups would be following different courses. On educational grounds, the parallel classes fit better the assessment purpose of driving student learning, but students seeking high marks might ask why they are being prevented from taking the more basic course.

As with many of the skills and abilities in this section, computing should not be seen in isolation. Once students have the skills, it is vital that they continue to use them. It follows that in theory modules, for example quantum mechanics, there should be assignments that rely on students using those skills, in just the same way as they will use the mathematics of matrices and partial differential equations. This is an example of *embedding* skills into teaching and assessment. For the proper acquisition of skills, two things need to happen. First, there needs to be some explicit training in the first place; it is not enough to tell a student to use a skill if they have not developed it. Second, they must continue to use the skills in the broader physics curriculum.

3.2.4 Ability to communicate

When I was working for my BSc in physics in the early seventies, I was expected to give an oral presentation in my final year but given no guidance on how to prepare for one. I was required to produce several (hand)written reports on laboratory work and projects for which I was offered some brief guidance but with little useful feedback on the finished article. These were not the good old days. More recently, partly as result of complaints from employers and partly due to a realisation that there is not much point in doing good science if one is unable to tell anyone about it, communication skills have become an important component of the training of a physicist. Indeed, in many countries that do not have English as a first language, the ability to communicate in English is a requirement. As before, it is not sufficient simply to offer students the opportunity to communicate, they must also be trained to communicate, which in turn has consequences for how the curriculum is designed and how the assessment is linked to that training.

Assessment

The types of assessment match closely the types of communication we wish to include. Right at the heart of good communication are the abilities to give a clear, interesting and informative talk and to explain ideas and concepts unambiguously in writing, with intelligent use of illustrations in both cases. However, there is a large range of tasks to which these two basic skills apply—typical assessments include: a live talk, a podcast, a blog or vlog, a newspaper or magazine article, a scientific paper; a review of a book or a paper. Some of these may be undertaken as a group.

Commentary

A dilemma in teaching and assessing communication skills is whether it should be done in isolation, as a separate course or module, or whether it should be entirely embedded in the physics curriculum. My view is that it is essential that students receive a good grounding on how to give a talk and how to write clearly and that training should be dedicated and specific to the task. However, it is also essential that the topics are based in physics (or whatever subject is under instruction.) In many universities, the traditional laboratory training sessions have been expanding to become more general skills training sessions, which include presentational aspects and, sometimes, computing. This, I think, is a positive trend because the structure of assessment for the acquisition of skills will be similar, whether we are considering experimental technique, computing or communication. The laboratory or workshop approach also lends a context to the presentations, in whatever form they take.

It is insufficient simply to require students to communicate; they must be provided with appropriate training. For technical points, such as how to present a podcast, some of that will be straightforward information. However, for the other elements, such as maintaining eye contact, identifying one's audience and choosing the right level, structuring the talk or article etc, there is a need for practice, formative assessment and feedback. Also, given the need for one presentation to build on another, the formative assessments need to be compulsory, which, in turn, requires them to have a summative element.

It is also essential that there is complete transparency in what is being assessed; i.e. that students are aware of what is important and that their assessment and feedback link to that guidance. Where students are giving oral presentations, I have found it useful to record the presentation so that the individual students can self-assess their own performance; these days, a perfectly good recording can be made on a smartphone or a tablet computer. It can also be helpful to introduce an element of peer review, getting the other students to rank their colleagues against the criteria. Not only does this help the speaker but it also helps the observer to identify both good and not-so-good practice. An example of a set of criteria for an oral presentation, which forms the basis for the feedback, is shown in table 3.1; the example is not without fault and will be returned to later in chapter 5. Note also that,

Table 3.1. Exemplar assessment guidelines for oral presentations.

Item	Outstanding	Good	Satisfactory	Poor
Introduction	Clear introduction with central theme and purpose stated.	Introduction gives the basic theme.	Introduction does not match talk or is too brief/long.	No introduction.
Science	Good, clear, imaginative explanations and error free.	No errors of science and consistent explanations.	No errors and adequate explanations.	Errors of science and/or explanations unconvincing.
Logical flow	Each section is logically coherent and links to other sections and the central theme. Good flow with evidence of planning.	Sections are logically self-consistent.	Generally argument is logical with perhaps a few discontinuities.	Haphazard and disorganised.
Quantity of content	Quantity is well matched to the time available. The talk does not sound rushed and does not have unnecessary detail.	Quantity fits the time available and is well-structured.	The talk is completed on time.	The quantity of material is inappropriate for the time available.
Matching to task	The talk matches perfectly the task in terms of content, level and identifying the target audience.	The target audience is judged correctly and the level well judged.	The talk matches the task and is reasonably linked to the target audience.	The talk misses the target.
Summary	It is clear that the talk has ended. There is a cogent summary of the key points. Where relevant, there is an indication of future work.	A clear summary is provided of the main conclusions which link to the points made.	There is a summary.	No summary.

Audience rapport	Plenty of eye contact with audience and positive body language. The speaker always looks at the audience when speaking to them. The audience is engaged. The speaker has no verbal or physical mannerisms.	Eye contact is maintained and the speaker has good posture, standing in a good position and largely avoiding verbal or physical mannerisms.	Eye contact generally maintained and no long periods looking away. Few tics.	No eye contact and poor body language. Words or phrases repeated and/or physical mannerisms.
Enthusiasm	The speaker is enthusiastic about the talk and the subject; they sound as if they care about the audience understanding.	Speaker seems to be interested in the material and the audience.	Speaker maintains an interest in the material and links to the audience.	The speaker seems to wish s/he were somewhere else.
Visual aids	Visual aids are clear, informative and well-chosen. Slides are not cluttered, make good use of colour and are clearly visible to the whole audience. They add to the talk.	Visual aids are clear and well-chosen. They are visible to all and relevant to the material.	Visual aids are clear and visible to all.	Visual aids are missing, unhelpful, irrelevant or cannot be seen clearly.
Voice modulation	Both the tone and the volume of the voice are used for emphasis. There is a variation of pace to avoid monotony.	The talk is delivered clearly with some modulation of voice, tone and pace.	The talk is delivered clearly with occasional modulation of volume, tone or pace.	Largely a monotonous tone with no variation.
Question handling	A positive response to questions, questions are answered with confidence and authority. Answers match the question are at the right level.	Questions are welcomed and answered accurately with reference to the talk.	Questions are answered correctly.	Questions are not answered and/or not taken seriously.

while the assessment has the primary purpose of the development of skills, students should always be expected to produce correct physics!

As a final comment, as with other skills, once the basic training has been completed, in going forward we should respect two principles. First, the skills should be embedded as part of the assessment for at least some of the other modules; projects are an ideal vehicle for any or all of the presentation formats but theory courses can also involve student-driven activities for which some kind of presentation is the output. Second, where they are embedded, the assessment includes some component for the presentational aspects as well as for the content. Thus, in marking a project report, the quality of the language is one of the marking criteria.

3.2.5 Conceptual analysis

I thought long and hard before including this skill in the list; there is a small step between conceptual analysis and conceptual understanding and I have already warned against the dangers of citing 'understanding' as a learning outcome. Philosophers have invested much time and effort in trying to explain what we mean by the term understanding; humble assessors should stick to measuring what students can actually do and not refer to any internal, mental processes.

By conceptual analysis of a problem or situation, I mean a qualitative explanation of what is happening as opposed to a purely mathematical one. To take a specific example, borrowed from Lillian McDermott [4]: consider some dc electrical circuits comprising batteries, resistors, light bulbs, switches etc, such as that shown in figure 3.1.

One can analyse such circuits by the simple application of Kirchoff's laws, provided one knows the values of the components and one is careful about signs and so on. But it is a quite different skill to be able to explain qualitatively what will happen, for example, to the brightness of each of the three bulbs, A, B and C, when the switch is closed without recourse to the equations. This is what I mean by conceptual analysis.

Stretching the definition a shade, I also include under this heading all the paraphernalia usually labelled as 'thinking like a physicist', which includes conceptual analysis but also includes back-of-the-envelope calculations, dimensional analysis, checking the plausibility of a conjecture or the result of a mathematical calculation, etc.

Assessment

Conceptual analysis may be introduced in a dedicated, first-year course but it should form part of all teaching within a physics degree programme. Of particular relevance is the inclusion of such analysis as part of the assessment of a laboratory course. This will be discussed in more detail in chapter 5 but it is easy to see how making estimates, checking the plausibility of measurements and designing investigations all lend themselves to the laboratory environment.

Within theory modules, multiple choice questions and short-answer questions are excellent probes of the ability to carry out conceptual analysis and these can be

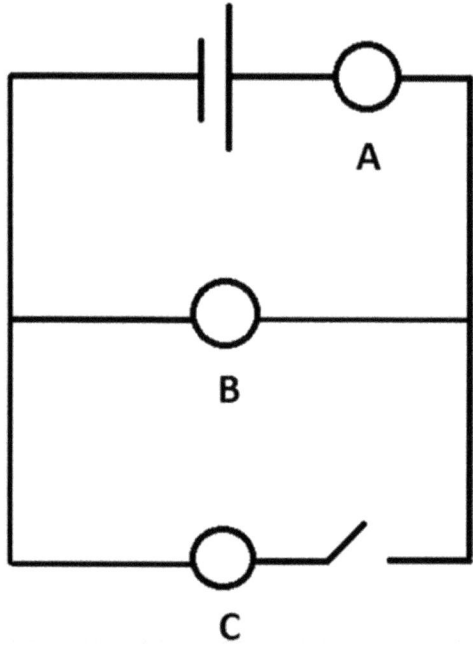

Figure 3.1. Simple electrical circuit of three bulbs and a battery.

incorporated into timed examinations, problem sheets and in-class tests. It is not so much the methodology of the assessment that matters in this case but the type of question asked. A typical example might be to ask why is that most solids expand as their temperature is increased and why the expansion coefficient disappears as the temperature approaches absolute zero. Note the open-ended character of the question, which allows a number of different, complementary explanations, depending on the level or the particular topic of the module.

The development of such skills is greatly enhanced by students talking to each other about the problems. So, on the principle that assessment drives student activity, arranging for students to work in small groups to answer such questions would be a valuable approach. As elsewhere, where problem sheets are part of the assessment, this group cooperation probably occurs naturally anyway, although there is virtue in making it more formal so that individual students are not excluded. In the latter case, it is good practice to vary the membership of the groups.

Commentary
In some areas of physics, mostly at the introductory levels in terms of university programmes, there exist standard sets of multiple choice questions which are given the generic name of *concept inventories* [5]. Most of these have been developed in the USA and improved over a period of time, although I have found that a few of the questions are ambiguous at best, reflecting the difficulty of reducing problems to a simple choice of four or five alternative answers. These inventories have been used

for many purposes, including research—some users have claimed that a few of the questions lead to significant differences in identifying the correct answers according to the gender of the student [6]. I have found the inventories most useful as formative assessment in helping teachers see the strengths and weaknesses of the students.

Most teachers would agree that conceptual analysis of the type I have described is an essential skill for a physicist, but, even where training in such skills is taken seriously, it usually occurs in the first or second year of a programme and is neglected later on. There is absolutely no reason why these types of skills cannot be tested as part of the assessment of specific topics, such as electromagnetism or quantum mechanics; the fact that they seldom are is because departments rarely take a coordinated view of assessment across a programme. I shall return to this point in chapter 6.

On a related point, conceptual analysis makes most sense in a synoptic context, since one of the most useful skills for a physicist is to decide which bits of physics apply in a particular situation. However, one does not generally find examinations on, say, quantum mechanics that require sophisticated use of ideas from electromagnetism. But real-life problems do require the physics to be identified and introducing assessments with a synoptic character, particularly in the later stages of degree programmes, helps in this respect, although I am not convinced that the artificial environment of a synoptic, timed examination offers the best methodology. A few universities, notably Leicester [7], but there are many others, have constructed a large fraction of their programmes around synoptic, problem-based learning.

3.2.6 Problem solving: logical reasoning and originality

Many people would say that the ability to solve problems is the defining feature of a physicist. For that reason, it is worth including as a separate item, although many of the other items appearing on the list link into problem solving: knowledge, mathematics, computing, information skills and, particularly, conceptual analysis. I have, however, included logical reasoning and originality under this heading as those virtues are central to this activity.

To be clear, by problem solving I do not focus on the problem sheets that are usually associated with specific topics covered in lectures and which almost always have unique answers, but on dealing with open-ended situations. The development of problem solving is best done in investigative environments, with student-led activities, and that will be explored further in chapter 5.

Problem solving is notoriously difficult to teach; in most cases, students are given projects, or problems to be solved with little guidance on strategies on how to proceed. It is beyond the scope of this book to discuss that issue except to say that the assessment, as ever, will drive student activity, so assessments should be carefully planned to link to the identified strategies and that means a series of staged assessments.

Assessment
Problem solving is rarely a solitary activity in real life. People work in teams and even people whose idea of joy is to sit alone in their office pondering an issue will

talk to others for inspiration and critical input. It follows that problem solving is best undertaken in small teams with the assessed outputs as one or more of short reports, oral presentations and/or posters, depending on the size of the task, which may vary from a simple qualitative question to a full-blown group project.

Commentary
It is common in physics programmes for lecture courses to include problems given to students with the expectation that they will work on them alone, although it is accepted that there will be some sort of discussions within friendship groups. These illustrative problems are certainly worthwhile but they do not constitute any serious attempt either to develop problem solving skills or to assess them.

First, problems should be set to small groups that are not self-selecting. Second, the problem solving process follows a number of stages; the exact details can vary but usually there is something similar to:
- Identify the problem—what are we trying to do and what output do we expect?
- What is the basic physics involved?
- What do we know about that and what do we need to find out about?
- Allocating tasks, e.g.:
    ○ Information gathering (people, library).
    ○ Calculations.
    ○ Computer programs.
- Synthesis.

The list has a degree of arbitrariness and often requires several iterations but what is essential in the assessment is that, particularly when they first encounter the activity, students provide evidence of having progressed through the defined stages.

The best exercises are synoptic in character, where students do need to show originality of thinking. A question here is what we mean by originality; for someone in the early years of a physics programme, it is impossible to display originality in the sense of new physics. What we mean by originality is new thinking for *them*. It follows that exercises should be chosen to allow that to happen, preferably without a set answer and preferably with the requirement for students to make choices about which directions to take. The point here is that the quality of the assessment is not determined by the level of the physics but by the student experience and their ability to make contributions. There is no point in setting problems that are beyond their capabilities.

As an example, a few years ago, I set a group project on how school physics examination papers have altered over the last 50 years. The project required a deep understanding of basic physics but the main steps involved: the collection of the papers; organisational decisions such as whether to look at topics or papers; the decision on which parameters to look at (e.g. difficulty, mathematical content, memory questions etc); the quantification of each of those parameters; the analysis of the examination papers, leading to possible iterations of the earlier stages;

synthesis and conclusions. Within all that, there was the allocation of duties within the team.

Just as the best set problems require synoptic thinking and decision making, they should also make use of other skills, such as computing, mathematics and communication.

3.2.7 Scientific methodology

By scientific methodology, I do not mean the details of experimental techniques but the interplay between theory and experiment, the way in which knowledge is accumulated. In essence, we have theory as a descriptor and, crucially, a predictor of experimental results. This process is unique to science; indeed, it is what defines science as a philosophy. The word *theory* is used regularly elsewhere, for example in literature and in some of the social sciences but generally in a different manner. The subservience of theory to experiment is at the heart of physics.

In my experience, most institutions expect students to acquire their understanding of how the scientific process works by a process of osmosis. Sometimes, historical examples of the interplay of theory and experiment are provided in lectures, such as the discovery of the neutrino or the Michelson–Morley experiment, but these are not usually analysed in terms of the scientific process. That process is best taught and assessed within a laboratory environment and the latter will be included as part of chapter 5.

3.2.8 Experimental competence

All physics programmes, even those with an emphasis on theory, require at least some appreciation of experimental technique. The topic is covered in detail in chapter 5.

3.2.9 Ethics and professional behaviour

One can argue that ethics and professional behaviour should be part of any degree programme in any subject. In science subjects, there have been many cases of falsification of data in the research laboratory, many of them high profile, as well as cases of plagiarism. More generally, there are issues to do with: intellectual property (IP), including patents; health and safety; diversity and inclusion; data security; and various legal constraints, such as protocols for working with people and other animals (an example of a statement on ethical behaviour is provided by the American Physical Society in [8]).

In a first degree in physics, the most relevant skills are those associated with honesty, interacting with colleagues, plagiarism, health and safety, and data security. IP and other legal issues are probably best left for research degrees.

Assessment
Of all the skills on this list, this one lends itself best to a benchmark assessment. Specifically, health and safety, data security, and diversity and inclusion are best dealt with via a code of conduct/guidelines approach, in which students are given a

short, unseen, knowledge test which they must pass, at whatever level is appropriate, in order to progress; they may need more than one attempt. Alternatively, the benchmark test could be given some flat credit, say 5% of a module, on a simple pass/fail. There is more discussion on how to incorporate benchmark assessments into an overall mark in later chapters.

Plagiarism is best considered within the communication skills assessments, with instructions for adequate referencing and what is and what is not allowed in terms of working with other students, as well as within the assessments for projects, which might include literature reviews and research-type papers. It is necessary, of course, for those instructions to be common to all relevant assessments across the entire programme.

Commentary

Plagiarism poses a special problem for assessment in physics in activities such as essays and stand-alone literature reviews: it is very difficult for an undergraduate student to say anything original about research-level science, so the best we can hope for is some sort of synthesis of ideas rather than a critique. For that reason, I am not keen on such assessments but, if they are used, they need to be carefully selected so that the student is able to assimilate the material using the physics they already know.

Having a code of conduct, of which students are required to demonstrate an awareness, is generally useful as it sets out from the beginning the ethos of the department or school and it means that students are thoroughly aware of what the responsibilities are of them and their teachers (a typical code of conduct, this one for the University of Cambridge, can be found in [9]).

3.2.10 Independent learning

Although working with others and talking about physics are excellent ways to improve understanding, in the end, learning is an individual activity and one emerges with an individual qualification. It is certainly true that traditional, unseen examinations operate at the individual level although, if we are not very careful, the independent learning they promote is primarily rote memorising and cramming in the last few days before the examination. In terms of life skills, whether in physics or elsewhere, those activities have limited applicability.

What we should mean by independent learning is for students to be able to acquire new knowledge and techniques for themselves. These days, the Internet is the primary source of information but does not provide much of a route to learning more substantial material or technical skills.

Assessment

Extended projects are clearly useful as means of developing independent learning but, in many cases, students are expected to have the skills already before embarking on the project. In some universities, a module or short course on study skills is provided, often supplemented by personal tutorials. In these cases, incorporating a benchmark pass/fail assessment, based on a specific task, would be a useful contribution.

Oral presentations also provide excellent opportunities for students to display their independent learning. Where communication skills are explicitly taught in stand-alone sessions, one of the criteria for assessment should include the content as acquired by the speaker or writer.

Commentary
In my experience, lifelong learning skills are not often included explicitly in the curriculum for a physics degree. Anyone who has taught will know that the process of preparing material and imparting it leads to a far deeper understanding that one can ever acquire by sitting in lectures. It follows that the best assessments to drive that type of learning is to require the student to communicate the results of their learning. That may be done orally or in writing but the former allows the possibility of questions and is therefore superior if it can be accommodated.

One radical means of inculcating learning skills is to employ a competence-led approach to learning and assessment usually called *mastery learning* [10]. This type of teaching is usually delivered using a flipped-classroom in which the material is learned by students in their own time with scheduled tutorials to cover key issues. Students progress though the material stage by stage and are only allowed to progress to the next stage once they have achieved competence in the preceding one. Their achievement is then measured not by the accumulation of marks but by how far they progress, with a bare pass defined by the minimum of material required for subsequent work. Although I have taught a course in such a way, it was used as an optional supplementary to the lectures and therefore not taken seriously by the students.

An interesting exercise in of using mastery learning has been described by Francis and Savage [11]. They used it to teach a large class and found that it worked very well, with students working harder and engaging well with the material. However, they deemed the exercise a failure on the grounds that the students were no better at answering examination questions than those who learned the material via traditional lectures. The assumption they are making here is that the timed examination is the true test of learning. I return to that assumption in chapter 4. For the moment, I leave readers to ponder its validity.

As a final point on independent learning, one critical skill for life, and for physicists above all, is to find the right person to talk to, both in terms of cutting out unnecessary work and for finding the right places to look for useful information. Unfortunately, the way educational systems work, this skill is not only undervalued, in some cases, it might be construed as cheating. The issue is that many of the tasks we set our students are those where the outcome is known and we would not consider it laudable for them simply to find a person who knows and copy the solution, particularly so if that person is another student. It is difficult to see how assessment can be useful in changing that situation, except by setting as many open-ended tasks as possible.

3.2.11 Team working

Anyone in any sort of employment is going to have to work as part of a team. Since teamwork poses special problems for assessment, I devote the next section to it.

3.3 Assessment of teamwork

There is no doubt that working as part of a team is an important graduate skill, whether one's future lies in or out of academia. When employers are asked about how well physics graduates make the transition into the world of work, they usually make positive noises but report two areas where there is room for improvement. The first is the ability to communicate well in a variety of formats to a range of audiences. The second is being able to work well with other people as part of a coordinated team. As a consequence, most UK physics degree programmes now include some teamwork. Although some academics may see such devices as diluting the content of a physics degree, in my view there is considerable educational gain even with regard to the physics material. First, students' understanding of physics increases by talking to each other about it. Second, weaker, or perhaps lazier, students raise their level of achievement, probably because the team environment improves their working methods.

Team working usually occurs as some sort of project work, possibly in conjunction with external sponsors, but it is possible to arrange more modest and imaginative team tasks, for example arranging for a group to teach a particular topic within a theory module. The methodology for assessing team projects is broadly the same as for individual or paired projects—plans, oral presentations, reports etc.— and these will be discussed later. However, a team project is most certainly not just a set of people, each pursuing an independent line of work; the effectiveness of a team depends not only on the quality of the individual work but how it is managed and brought together. Indeed, those aspects present significant challenges to assessment, keeping in mind the purposes we require of it; specifically:

- The team aspect of the exercise must be recognised. It is not sufficient solely to assess the individual contributions, there must be some recognition of the overall achievement of the group. I have known individual students who have withheld important information from the other team members because they wanted to be seen to be making more progress. While that student might well have been doing excellent work, such selfishness should not be incentivised.
- Effective team working does not mean everyone doing the same thing. Not everyone can be a leader and there is as much need for the quiet person who carries out careful analysis as there is for the person who comes up with lots of ideas. However, that means that each student will be responsible for a different task, which means the learning outcomes, and hence the marking criteria, might be different, leading to concerns about consistency.
- Even leaving aside the point above about the different roles, it may also be the case that one or more students does not pull their weight. In one sense, that is an issue for the team itself to sort out but, in a real environment, there would be some management process to prevent such a situation. In an educational context, where students are concerned about being dragged down by others, it can lead to conflict and student perceptions that they are being unfairly treated.

It is clear, then, that assessment of teamwork requires specific components to overcome the issues mentioned above. I suggest the following.

3.3.1 Team assessment

The assessment must include elements that refer to the team performance, since the ability to work together is the skill being tested. However this is achieved, it is essential that the organisational aspects are in place at a very early stage in the project. Therefore, it is a good idea to award a small number of marks for specific items such as the allocation of roles and the agenda and minutes for formal meetings. By recognising these aspects by awarding credit, one is indicating their importance.

Most projects have the largest proportion of the marks allocated to a final presentation, written and/or oral (or equivalents such as websites, podcasts or even artefacts, such as a computer program.) These elements should also have a team mark as well as marks for individuals (see below). The criteria for the team mark can include organisational aspects but, mainly, they can be essentially identical to those for individual projects, as discussed in chapter 5. In essence, one is assessing the achievement of the team by looking at the overall progress and, even though it may occasionally be the case that much of the progress is due to one individual, or that the project was severely hampered by someone, the team mark should still reflect the overall progress.

Although the team mark is the same for all members, there will usually also be marks given to individual contributions, so the question arises: what is the split? I suggest that 50/50 works well for large projects, since it indicates what the exercise is about. For example, awarding only 10% to the team aspects sends entirely the wrong message and a figure of 90% would exacerbate student concerns about fairness. However, if the groupwork is a small component of a module, maybe just a few per cent, in parallel with a much larger assessment, perhaps an examination, it is not efficient to introduce a complicated machinery for separating individual and team aspects; in those circumstances, it is almost certainly best just to award a team mark.

3.3.2 Peer and self-assessment

If the team works well the teacher will not directly witness most of what is going on and, in particular, the dynamics of the group and the weight of the various contributions. For the benefit of the assessor and also to allow the students some input into the process, it is a good idea to require the students to comment on the contributions of their colleagues, as well as to give a statement on their own activity. This can be done in several ways but here are a few ideas:
- Each student can allocate a % contribution for each of their colleagues, both in terms of their individual work and also in terms of their contribution to the team. A possibility is that each student contributes two such breakdowns, one just including their colleagues and the other for everyone, including themselves.

Table 3.2. Sample peer feedback form

Name	
Task	Write the name of each of your group members in a separate column. For each person, indicate the extent to which you agree with the statement on the left, using a scale of 1–4 (1 = strongly disagree; 2 = disagree; 3 = agree; 4 = strongly agree). Total the numbers in each column.

Evaluation criteria	Group member	Group member	Group member	Group member
Attends group meetings regularly and arrives on time.				
Contributes meaningfully to group discussions.				
Completes group assignments on time. Prepares work in a quality manner.				
Demonstrates a cooperative and supportive attitude.				
Contributes significantly to the success of the project.				
Totals				

- Each student provides a brief statement—say a page or so—on what their personal contributions were and how they contributed to the team effort.
- They can then also make similar, but briefer, statements for each of their colleagues. To simplify the latter step, it is possible to provide a template, such as the one shown in table 3.2, but I have found that no standard template is entirely appropriate to cover all projects.

It is a separate question as to what use is made of this student input. I have tried in the past to allocate a team mark then use the student percentage allocations to provide discrimination within the teamwork mark. This approach does have the great advantage of rewarding those students who really do commit to the team ethos but it loses the benefit of everyone being responsible for the team's success. Perhaps a better approach is to include this input as a small component of the individual mark for a student. A third possibility is not to use it in any formal way but that makes it hard to justify and we would not wish to lose the benefit to the students in reflecting on the various contributions. Some students baulk at having to rate their colleagues in terms of giving them a numerical mark, so a qualitative response may be preferable.

It is possible for peer assessment to contribute to the final, summative mark. Students that do engage with the process gain a great deal in terms of their understanding of assessment but, in practice it does not work well and is too unreliable [12]. Students themselves do not like to do it but the overriding issue is one of consistency.

3.3.3 Individual assessment

The form of the assessment for a group project can be similar to that for any project (see chapter 5), comprising components such as a project plan, written report and oral presentation(s). To maintain the team approach, some elements (for example, the project plan and the minutes of meetings etc) should have a single assessment for the team. Others should have individual components but should not be a set of independent reports or talks. There should be a single project report and/or presentation, marked according to whatever criteria are defined, but it has to be clear which student was responsible for which section. Each student, as part of their statement, should state precisely what contribution they made and how it fitted into the overall work of the project. Note that even the individual mark can, and probably should, carry a component for the contribution that individual has made to the team; the determination that component can take into account the peer assessment forms.

3.3.4 Clarity and transparency

For most students a team project will be a novel activity and there will naturally be concerns about what is expected and how it will be assessed. On the latter, as ever, clarity and transparency are paramount and all material pertaining to the assessment should be provided to students at the outset of the project, including specimen assessment forms and a full breakdown of how marks are allocated, including the teamwork elements, and an indication of the criteria used for marking. A possible breakdown of marks is given in table 3.3. I am aware that some will find the breakdown unnecessarily fragmented. My justifications are that, for clarity and consistency, it is best to be as explicit as possible about what is being assessed and that, by giving credit for an item, one is demonstrating its importance.

To finish this section, it is worth making a couple of general comments. First, it is common for the spread of marks in any project, even more so for a team project, to be lower than for a typical, timed examination. There are several reasons for this and one is that it is psychologically harder for an assessor to give a fail mark explicitly for, say, a talk, as opposed to adding up a set of marks for an exam paper and ending up with a fail mark. A similar comment might apply for very high marks. But it is also the case that the lower-achieving students do improve the quality of their output in groupwork, partly because the group discipline is imposed on them and partly because they learn from, and are motivated by, discussions with their fellow students. The absence of very low marks is genuine and not an artefact of the assessment.

Table 3.3. Example of a possible mark allocation for a team project.

	Team	Individual
Project plan	10%	0%
Agenda/minutes/allocation	5%	0%
Interim report	10%	0%
Final report	20%	30%
Oral presentation	5%	10%
Contribution to team	0%	10%
Total	50%	50%

The second comment is that groupwork always presents the prospect of one of the team not pulling their weight, either due to illness or for some other reason. That does not usually need to cause a major issue with the assessment; any team has to cope with this type of eventuality and credit can be awarded for how the team responds to such setbacks. However, it is a good idea to build in a confidential, whistleblowing facility to allow one or more members of the team to indicate a problem. Such a mechanism can be very useful if a member of the team is feeling excluded and it is as well to pick up such issues as early as possible.

3.4 Final remarks

I obtained my BSc degree in physics from the University of Birmingham in 1973. For most of the programme, we had a diet of lecture courses of variable length (no modules in those days!) each assessed by a final unseen examination. There were laboratory sessions and these were assessed principally on the basis of handwritten formal accounts, with the odd interview. We received little or no feedback on progress and were not even told our marks for the papers, just our final degree classification.

This experience was more or less common across UK universities at the time. What was not common, however, was that following the final examinations, at Birmingham, we undertook a six-week 'group studies' workshop, during which we wrote essays, gave talks, solved problems and undertook a major experimental project [13]. The exercise, which was developed by Paul Black and others, changed completely my view of physics; during those six weeks, I worked harder than at any time during my university career to that point, enjoyed myself and, perhaps for the first time, had a feeling for what physics was all about. I was immersed in physics instead of just learning about it. This experience demonstrated to me that learning physics is far more than just sitting in lectures and has informed my approach to education throughout my academic career. I know that other Birmingham graduates have been similarly energised.

Happily, contemporary degree programmes pay far more attention to 'what it is to be a physicist' than in my day. In this chapter, I have tried to provide a comprehensive list of the properties we associate with graduate physicists and to try to indicate how those elements may be assessed with the framework of available methods and satisfying our principles of assessment. However, since this book is very much intended to be a practical guide, I have to mention that a substantial fraction of academics are still highly sceptical of assessments that are not traditional examinations, perhaps because that is the experience they had and they feel it did them no harm.

Such people are not easily persuaded but I do think it helps to begin the conversation by looking carefully, as in this chapter, at what we want our graduates to be able to do when they leave. From that we can decide how those things can be taught and develop learning outcomes for each module (or component course for those who do not work in modular systems). The assessment then links to the learning outcomes in the ways discussed above. The main point conveyed, I hope, is that such a range of desirable outcomes requires a diversity of assessments. While there is undoubtedly a role for formal examinations, it is nonetheless the case that assessments that lead to more active student participation and that encourage students to think and talk about the subject, drive more efficient student learning and, ultimately, produce higher levels of student satisfaction. That was certainly the case for me at Birmingham.

Bibliography

[1] Institute of Physics 2021 Degree accreditation and recognition *IOP*
[2] QAA 2019 Subject Benchmark Statement for Physics, Astronomy and Astrophysics *The Quality Assurance Agency for Higher Education* https://qaa.ac.uk/docs/qaa/subject-benchmark-statements/subject-benchmark-statement-physics-astronomy-and-astrophysics.pdf
[3] Garner J and Denny P 2019 Mastery learning in computer science education *Proc. of the Twenty-First Australasian Computing Education Conf.* pp 37–46
[4] McDermott L C and Shaffer P S 1992 Research as a guide for curriculum development: an example from introductory electricity, part I: investigation of student understanding *Am. J. Phys.* **60** 994–1003
[5] Laverty J T and Caballero M D 2018 Analysis of the most common concept inventories in physics: what are we assessing? *Phys. Rev. Phys. Educ. Res.* **14** 010123
[6] Madsen A, McKagan S B and Sayre E C 2013 Gender gap on concept inventories in physics: what is consistent, what is inconsistent, and what factors influence the gap? *Phys. Rev. ST Phys. Educ. Res.* **9** 020121
[7] Raine D J 2019 *Problem-Based Approaches to Physics* (Bristol: IOP Publishing)
[8] The American Physical Society 2019 Guidelines on Ethics (Full Statement) *APS* https://aps.org/policy/statements/guidlinesethics.cfm
[9] Ward D 2008 Plagiarism, Collaboration and Cheating *Department of Physics, Cambridge University* https://www.phy.cam.ac.uk/students/teaching/resources-links/plagiarism
[10] Keller F S 1968 Good-bye teacher *J. Appl. Behav. Anal.* **1** 79–89
[11] Francis P and Savage C 2009 Mastery learning in a large first year physics class *Proc. of UniServe Science Conf.* pp 152–9
[12] Wilson M J, Ming Ming D and Huang L 2015 I'm not here to to learn how to mark someone else's stuff *Assess. Eval. High. Educ.* **40** 15–32
[13] Black P J, Dyson N A and O'Connor D A 1968 Group studies *Phys. Educ.* **3** 289

IOP Publishing

Assessment in University Physics Education

Peter Main

Chapter 4

Timed examinations

4.1 Introduction

In the previous chapters I explored the purposes of assessment, the methods of assessment and what we are trying to assess. In this chapter I will concentrate on timed examinations, not because they necessarily represent the best form of assessment but because they remain the most common type in physics degree programmes. In the papers presented, I shall look at the questions in detail and examine what the question is requiring of the student, not only in terms of the specific answer but also in terms of what would constitute their best strategy for preparing for the examination paper. I should say that the papers were chosen for their typicality, not because they were particularly well or poorly set.

In terms of the analysis, a simple inspection of the paper is not always diagnostic in terms of determining what is required of the student. As we shall see, there are many cases where a relatively complicated calculation or manipulation of ideas is requested and it makes a world of difference to both the level of difficulty of the question and how the student prepares for the paper whether that topic has been covered in detail in the lectures or if the students are seeing the question cold. Even model answers do not always reveal the degree to which a question is essentially bookwork or a genuinely new problem. Many external examiners, including me, insist on model answers indicating what each element of the question requires, for example bookwork, unseen problem, synthesis of ideas etc, but the only people who really know are the people teaching the course (who are usually also setting and marking the paper) and the students themselves.

One skill which, in principle, could be tested by timed examinations is the quality of written communication. However, in practice, it rarely is; the hard-pressed markers of a descriptive section generally have a list of key points that, if mentioned, accrue the marks. Whether the points are made in coherent, unambiguous sentences is not commonly taken into account. Personally, I feel that a lost opportunity but there is the reasonable justification that students working under examination conditions do not have time to write polished sentences. Either way, if students are to be required to express themselves clearly in timed examinations, they must be told of the fact.

Examples of the criteria used for assessment of student-centred activity, together with the assessment of laboratory and project work are considered separately in chapter 5.

4.2 Structure and nomenclature

4.2.1 Structure

Formal examination papers typically last 2 or 3 h; any shorter than that and the reading time for the paper becomes a too significant fraction of the total. There is a large variety of structures of examination papers, sometimes even within the same institution. The simplest is just a set of questions of equal weight and the candidate chooses a subset, say 3 out of 4 or 5, to answer. This structure is common in the later years of degree programmes. The argument for the choice is that it enables a student to demonstrate depth in their learning. The disadvantage is that what one gains in depth one loses in breadth and the element of choice encourages students to 'question-spot', which can lead to them missing out whole sections of the course material in their preparation for the examination.

More commonly, and particularly in the earlier years, examination papers are split into two or more sections. A typical model is for there to be a section A and section B, with the latter having a structure similar to the one described above, with students attempting a subset of a number of longer questions. Section A often involves a set of short-answer questions, which allows a wider selection of the syllabus to be covered, albeit at a more superficial level.

There are some interesting variations within the section A/section B model. Usually, students are able, or required, to attempt all the questions in section A, but the total marks for that section may be capped. For example, there might be six questions of 5 marks each, with a maximum mark of 20 for the section. The reasoning for this approach is based on the observation that long, section B type questions tend to lead to a large range of marks; students who know the topic of the question well, can be awarded full marks whereas someone who gets off on the wrong foot may fail the question badly. In other words, the long questions can penalise students as much for what they do not know as reward them for what they do and, with only a couple of questions on the paper, if a student struggles with one question, it often spells trouble for the whole paper. Section A allows students to demonstrate what they do know across a broader range. It is often the case that section A questions are more straightforward than those in section B.

Section A, in the way described above, acts as a *compressor* on the range of marks. The most able students get full marks by answering, in the example given, all six questions to a high standard but are awarded 20 marks. Other students might answer some of the questions and accumulate, say, 18 marks, which they will retain in full; a difference of 12 marks is reduced to 2.

This compressor, which we introduced in Nottingham in the 1980s and which is now commonplace, is interesting in two ways. First, it is a rare example where the sanctity of the marks is disregarded. Of course, students are fully aware that section A counts only for 20 of the 30 marks available so there is no problem with transparency, but, nonetheless, for many students, not all the marks are counted. Perhaps surprisingly, students accept the arrangement, indicating that other creative approaches to the accumulation of credit may also be acceptable. Second, the compressor is a move in the direction of a benchmark assessment, in which a student is given a pass/fail. In terms of its approach, section A allows all students to pass the module by showing what they know and section B stretches them by asking questions to allow candidates to demonstrate what they can do in specialised topics.

Another model for section A is the use of multiple choice questions, particularly for the earlier years. These are useful for testing conceptual analysis and are often used as formative assessment within teaching sessions using technology to produce real-time results. If included as part of summative assessment, there is no reason why they should not be used as a compressor in the same way as for the short questions discussed above.

One aspect of the structure of papers now universally applied is that students are told how many marks are available for each sub-section of a question on a paper. In the artificial environment of an examination, this is a useful development since it indicates to students how much time they should spend on that part. In some cases, there are also explicit recommendations on how long students should spend on a question or a section.

4.2.2 Nomenclature

Any perusal of a set of physics examination papers immediately reveals a sort of formal exam jargon, which is the setter's attempt to be as unambiguous as possible in stating what the students are expected to do. Within that jargon, there are indicator words that signal the type of answer is required. I have never come across any lexicon that tries to define what these various terms mean in a coherent manner; what follows is my attempt for some of the more common terms.

State/name/write down/give: these terms require a simple statement, where the candidate writes something down in a few words. There is no need to justify what is written down. Sometimes algebraic expressions or numerical values are required.
Examples:
- State Newton's second law of motion.
- Name the thermodynamic quantity conserved in a Joule–Kelvin expansion.

Define: generally, state what a named quantity is in terms of simple physical ideas or quantities. While this sounds straightforward enough, definitions of quantities are not universally agreed and can also be different in different modules. In thermodynamics, entropy can be defined in terms of reversible processes, whereas in statistical physics the definition is usually couched in terms of the microstates of the system. In practice, *define* means to write down the definition that was provided in the notes for

a physical quantity. The definition may be all in words or can involve an equation. Note in the examples below, there is an ambiguity as to whether what is required is the verbal definition or a defining equation.
Examples:
- Define the Schwartzschild radius of a black hole.
- Define the expectation value of a physical variable.

Discuss: the invitation to 'discuss' in an examination paper is one of the more vague instructions. The word is common in essay-based subjects ('Discuss the implications of the Enclosure Acts on the British class system.') but, in physics, the meaning is often less clear. Sometimes it is used as a synonym for *explain* but there is also another common usage, which is to be avoided and which is to request something like: 'Discuss the significance of the result (of a calculation)'. Unless the examiner is asking the candidate to reproduce a remark from their notes (which in itself is of dubious value), such a request is vague and easy to misunderstand unless the context of the discussion is crystal clear.
Examples:
- Discuss how considerations of the symmetry of the wavefunctions of particles leads to their classification into fermions and bosons.
- Calculate the gravitational force between the electron and proton in a hydrogen atom and discuss the significance of your result. (*An example of the type to be avoided; why not just ask them to compare with the electrostatic force, if that is what is intended?*).

Explain/What is meant by?: indicates a mainly qualitative description of some relationship or process. We might have: *explain why* or *explain how*. A problem with using the word is that it is not always clear what the student can take as given in terms of the explanation or the level of detail required. For example, a question that asks candidates to 'Explain the origin of the Greenhouse Effect.' could simply mean a qualitative description in terms of absorption and re-emission of electromagnetic radiation or it might want a more in depth explanation involving absorption lines, equations of thermal balance and others gases as well as CO_2; in principle, the question could even be interpreted—by a geographer perhaps—in terms of social change, although that would be atypical in a physics examination. This issue can be overcome to some extent by offering a hint, as in the first example below, or by the number of marks allocated.
Examples:
- Use band theory to explain the difference between metals, insulators and semiconductors.
- Explain why the typical half-life for an α-emission process is larger than that for β-emission.

Derive: usually an instruction to follow some mathematical process, including physical reasoning, to produce the required equation. One essential requirement of such an instruction is that the starting point for the derivation must be absolutely

clear to the students. Woolly statements such as 'starting from first principles' are to be avoided. In some cases, the derivation will be a standard one, taken directly from the course notes, in which case, unfortunately, it is not uncommon for the starting point to be omitted. A frequent addition to this type of question is 'stating clearly any assumptions made', which usually indicates that the derivation is indeed one that the students have seen before.
Examples:
- Assuming that Hund's rules apply, derive an expression for the spectroscopic terms of the ground state of chlorine.
- Using the covariant derivative and the Lagrangian for fermions, derive the terms in the Standard Model Lagrangian which describe the couplings between the W-bosons, $W_\mu^\pm$, the τ-lepton and the τ-neutrino.

Verify/show that/prove: this instruction is similar in many ways to *derive*, in that the candidate is given a starting point and asked to derive an expression or prove some statement. The important difference is that, in this case, the result required is given in the question. As with *derive*, it is important to be clear about the starting point for the verification or proof. It is almost always a bad idea to request a candidate to show that a parameter has a particular numerical value. Given that many students are not always conscientious about explaining their reasoning, one is likely to get a page of numerical calculations with few words of explanation, ending with the statement that the parameter has the numerical value required. Marking is a difficult enough task without having to unravel such a mess.
Examples:
- If H and K are subgroups of G, prove that the intersection $\mathbf{H} \cap \mathbf{K}$ is also a group.
- Starting from the relativistic expression for the total energy of a particle of rest mass, m, show that for $v \ll c$, the kinetic energy is approximately $\frac{1}{2}mv^2$.

Calculate/evaluate/determine/find/deduce/infer: one of the more straightforward instructions to use the information provided to produce an algebraic or numerical answer. It is often a good idea to qualify the instruction with a phrase that calls for the logic of the calculation to be spelled out, both to help in giving part marks if the calculation is not completely correct, and to make the students demonstrate their thinking process.
Examples:
- Evaluate the commutator $[\hat{x}^2, \hat{p}_x]$.
- Calculate, in the centre-of-mass frame, the minimum kinetic energy protons must have in the head-on collision $p + p \rightarrow \pi^+ + d$ to produce the deuteron d and the pion π^+.

(Obtain an) estimate: an instruction that plays the same role as *calculate* but with an indication either that the answer is intrinsically approximate or that some assumptions have to be made. Sometimes, those assumptions are already implicit in the way that a question has been set. For example, any calculation using first order perturbation

theory is necessarily approximate. On other occasions, the candidate is expected to introduce assumptions or approximations in the calculation itself; in those cases, it is good practice to require the students to state what the assumptions are.
Examples:
- Estimate how long it would take a nitrogen molecule of mass 5×10^{-26} kg to random-walk one metre. (*An example of intrinsic approximation.*)
- Assuming a free electron model, obtain an estimate for the Fermi energy for gold, which has an electron density 5.90×10^{28} m^{-3}.

Solve: a very specific and straightforward instruction to solve a given equation, usually given a particular set of circumstances.
Example:
- Solve the first order ordinary differential equation

$$\frac{dy}{dx} = \frac{y+2}{2x+1}$$

with the boundary condition $y = 2$ at $x = 1$.

Draw/sketch/with the aid of diagrams: a straightforward instruction almost always accompanied with the requirement to label axes appropriately with the quantity and the scale.
Examples:
- Draw the *P–T* phase diagram of helium below 5 K. Label your axes, each of the phases and indicate any important points or lines.
- Sketch a typical hysteresis loop for a ferromagnetic material.

What? Which? How? etc: any examination paper has many examples of questions that are simply stated using this type of word. In the majority of cases, the meaning is clear, but in some cases while the examiner might be sure which answer they require, the meaning may not be so obvious to the candidates.
Examples:
- Why is Ni a ferromagnet, but Cu is not, even though they are neighbours in the periodic table?
- We want to completely ionize the helium, by first removing one electron then the second. Which of these electrons costs more energy to remove, the first or the second? (*It would be better to ask for some reasons.*)
- What conclusions do you draw about real systems that exhibit a tri-critical point? (*This example is taken from the end of a question so the lack of context is not a problem but it is nevertheless included as an example of poor practice. Taken literally, it is just asking candidates to state what they conclude, which is absurd, as there can be no correct answer.*)

Comment/justify/explain your answer/compare: these instructions often appear at the end of a question and almost always lead to ambiguity. Generally, they are a shorthand way of saying 'there is something in your notes on this point and I want you to write it down.' My advice is to avoid such types of question unless it is absolutely clear what the candidates are expected to say.

Examples (all poor practice):
- Explain your answer.
- Comment on the significance of your answer.
- Compare your result with the experimental value and comment on any difference.

4.2.3 Information for candidates

It is common for a section B style examination question to be concerned with one topic or a couple of related topics. A typical structure might start with a simple definition, following on with a proof or an explanation, then conclude with one or more calculations. Indeed, this single-topic structure is so engrained into the system that, if a question does not follow this template, it is important to flag that fact to the candidates in some way, either by sectionalising or providing hints.

Often, the structure of a question requires results from earlier parts to be used later. Where this is the case, it is important for a student who makes a mistake to be able to obtain credit for later work. For that reason, it is common for the early sections to be 'show that' questions, so that the result is given and can be used later.

The difference between an open-book and an unseen timed examination is that, in the latter, the only information that a candidate may draw upon is provided on the examination paper. It follows that, for those examinations, it is good practice for candidates to be made aware of what proofs, formulae etc, they are expected to remember. On the paper itself, a list of the values of the most important numerical constants (c, h, e, etc) are universally provided but it is also commonplace to include a list of useful formulae or data that might be required in answering the questions. Sometimes this list may be quite comprehensive and extend over a few pages, which rather blurs the distinction between unseen and open-book. However, whatever extra information is included on a paper must be provided to the students well in advance of the examination, so that they are familiar with it. If they are to be provided with equations, then it is as well that they are able to know how to find them.

Within the question itself, it is an obvious point to make but there must be sufficient information provided for a student to be able to answer it. As a rule of thumb, a student who has studied the material thoroughly but who is not necessarily conversant with the details of the lecture notes provided should be able to answer the question. While I hope that is seen as reasonable, it is a fiercer constraint than it first appears and, when analysing papers later in this chapter, I point out where I think it has not been followed.

4.3 Quality control

The quality control mechanisms in universities insist that examination papers are seen by people other than the setter. In essay-based subjects the setting of a question is relatively straightforward; it is the marking where subjectivity occurs and, for that reason, such papers are marked by two people independently. In physics, as with mathematics and other science subjects, this sort of double marking is rare, which means that there is a strong emphasis on getting the setting process correct. A typical scheme would be for the paper and its model answer to be checked by another academic, followed by some sort of moderation process to ensure that the challenge and time-requirement are consistent with those of the other papers at the same level. Finally, the papers are sent to one or more external examiners for their comments.

One might imagine that such an extended process would produce a polished set of examination papers of uniformly high-quality. Unfortunately, my own experience as an examiner, external examiner and head of department does not bear this out. Frequently, external examiners receive papers containing errors, in some cases rendering a question impossible, and it is far from unknown for a paper containing mistakes to be presented to students. More than once, candidates have been asked to prove an expression that was not correct. My belief is that much of the reason that this happens, when so many people are involved, is precisely because so many people are involved! Everyone thinks that someone else will look more carefully.

Given that anyone can make a mistake, the chief responsibility for avoiding errors lies with the academic checker. The role of that person should be to attempt the questions *before* looking at the model answers; all too often, the checker, undoubtedly pressed for time, glances through the questions, sees that they are reasonable and moves on.

I have also found that the setting process benefits hugely from having someone delegated to read through the papers to check that the language used is unambiguous and reasonably grammatical. Evidently, it is essential that the person doing this is a native speaker of whatever language is used. They can also consider some of the points referred to in the lexicon above.

4.4 Marking

Given that the topic of the book is assessment, one might have expected a great deal on marking but, in fact, in terms of traditional examinations in physics, marking is straightforward and all the work is done in creating model answers. However, I thought that it would be useful to offer some advice to novices in setting and marking.

- Always write out model answers in full, as you would expect an ideal student to write them, and make it clear what the marks are to be awarded for. This habit makes the marking easier and also helps gauge how long the question might take to answer.
- Indicate what type of exercise each section of a question is: bookwork, known problem, unseen problem, synthesis etc. Not only does this help the setter to balance the paper but it also helps external examiners judge what is being assessed.
- Mark one question at a time; do not try to mark the whole paper for each student sequentially. By doing them singly, you improve the consistency of your marking. Also, as often happens, you might realise some subtlety in the

marking, perhaps something you have been accepting that you decide is wrong, and it is much easier to go back to check the earlier scripts with this approach.
- Similarly, do not be afraid to make small changes to the marking scheme, provided that you do so consistently. The scheme on the paper is just a guide and, while wholesale changes would indicate a major problem with the setting/checking process, small changes are sensible.
- It is harder to mark consistently than it first appears. It is a human response that, if the first few answers produce low marks, one tends to become more generous and, equally, one becomes more particular if the first few marks are high. Obviously, such an effect can lead to unfairness; good practice is to mark, say, the first 20% of the scripts and then go back to skim over them to check if the marking is consistent. It is also good practice not to mark the scripts in the same order for each question—do not always start with the same person.

4.5 Analysis of examination papers

In analysing assessments, in the present case timed examinations, I comment on a number of features concerned with the structure, the clarity of what is being asked and how the papers link to learning outcomes or, if you prefer, to try to examine what student preparation would lead to them achieving a high mark on the paper. Examples are drawn from several universities but are presented anonymously. There is no attempt to compare levels between institutions, which would be impossible anyway without detailed knowledge of how the material was taught.

As part of the analysis, I describe the specific form of the elements in a question and will use the following terms:
- Bookwork: material that is reproduced from memory. Bookwork might include:
  - Definitions.
  - Derivations of formulae.
  - Rote-learned explanations.
- Seen problems: problems where the students have had the same, or very similar, problem as a worked example or on a problem sheet.
- Unseen problems: problems that the students have not previously encountered.
- Interpretations and explanations that go beyond bookwork.
- Other.

Since I do not have direct experience of the teaching that led up to these assessments, in some cases, it is not possible to distinguish whether students have seen problems before.

4.5.1 Year 1 paper

The first example is a Year 1 paper. As with all the examples I present it as the students saw it, editing only various bits of parochial rubric and, in some cases, removing extensive formula sheets. I suggest that the reader considers this paper in the light of what has been discussed earlier *before* reading any of my comments.

Fields, Waves and Matter Year 1
TWO hours
SECTION A
Answer as many parts of this section as you wish. The final mark for this section will be capped at 20.

(A.1) The threshold wavelength for photoemission of electrons from a sample of metallic potassium is 564 nm. Calculate the maximum speed of the photoelectrons that will be emitted when the metal is irradiated by light with a wavelength of 300 nm. [5 *marks*]

(A.2) The maximum wavelength of light that a typical silicon photocell can detect is 1.11 μm.
   (a) What is the energy gap (in eV) between the valence and conduction bands for the photocell?
   (b) Explain why pure silicon is opaque. [5 *marks*]

(A.3) Describe the main features of covalent bonds. Explain what distinguishes them from ionic bonds. State examples of systems that are held together by covalent bonds. [5 *marks*]

(A.4) Specify the ground state electronic configuration of an oxygen atom ($_8$O) and draw its orbital level diagram including the electronic occupation. List and describe the rules you applied to determine the ground state electronic configuration. [5 *marks*]

(A.5) The rotational constant of a linear rigid rotor is given by

$$B = \frac{h}{8\pi^2 c \vartheta},$$

where h is the Planck's constant, c is the speed of light and ϑ is the moment of inertia. In a CO molecule the equilibrium bond length is 0.1128 nm. Calculate the separation between two lines (in cm^{-1}) in the rotational spectrum of CO (the mass of carbon is 1.993×10^{-26} kg and the mass of oxygen is 2.656×10^{-26} kg).
   In which part of the electromagnetic spectrum does it lie? [5 *marks*]

(A.6) Consider the case where an electron e$^-$ and a positron e$^+$ annihilate each other and produce photons. Assume that these two particles collide head-on with equal but slow, speeds.
   (a) Show that it is not possible for only one photon to be produced.
   (b) Show that if only two photons are produced, they must travel in opposite directions and have equal energy. Calculate the wavelength of each of the photons. In what part of the electromagnetic spectrum do they lie? [5 *marks*]

SECTION B—Answer ONE question
(B2) Consider a particle of mass m that moves back and forth in a one-dimensional box of width L. Quantum mechanics predicts that the only allowed values of the

particle's energy are given by the equation: $E_n = \frac{h^2 n^2}{8mL^2}$, where h is Planck's constant and $n = 1, 2, 3, \ldots$.
  (a) An object of mass 1.00 mg is confined to move between two rigid walls separated by 1 cm. Treating this system as a particle in a one-dimensional box, calculate (i) the associated classical speed of the object in its ground state, and (ii) the excitation energy required to promote the object to the next available energy state when the speed of the object is 3 cm s^{-1}. Do the quantum mechanical results in (i) and (ii) agree with our everyday experiences? Explain your answer. [4 *marks*]
  (b) Consider an electron in the box. Determine the width L of the box for which the ground state energy is equal to the absolute value of the ground state energy of a hydrogen atom. Compare L to the size of the hydrogen atom. [3 *marks*]
  (c) For the states with $n = 1$, $n = 2$ and $n = 3$, sketch the shape of the wavefunctions and of the associated probability distributions. Explain why the results for the probability distributions contradict what would be expected for a particle that moves in the box according to Newton's laws. Under which circumstance would the classical and quantum mechanical results agree? Explain your answer. [5 *marks*]
  (d) Calculate the de Broglie wavelength of the particle in a one-dimesional box of size L for any value of n. Describe another physical system for which the same relation between a wavelength and a distance L holds. [5 *marks*]
  (e) State the Heisenberg uncertainty principle. Verify that this principle is satisfied for a particle confined in a one-dimensional box. Explain why the fact that the lowest energy of a particle confined in a box is not zero is a direct consequence of this principle. [7 *marks*]
  (f) Estimate the zero-point energy of an electron inside a one-dimensional box of length 10^{-14} m, the order of magnitude of nuclear dimensions. Compare it with both the gravitational potential energy and the Coulomb potential energy of an electron and a proton separated by the same distance. On the basis of this comparison, discuss the possibility that an electron can be confined to the nucleus. [6 *marks*]

(B3) Consider a carbon dioxide molecule (CO_2). This is a linear molecule in which the carbon atom is in the middle between the two oxygen atoms.
  (a) How many normal modes of vibration has the molecule? Sketch them. [4 *marks*]
  (b) Spectroscopic measurements show that CO_2 has one Raman active mode at a frequency of 1337 cm^{-1} and two infrared active modes at frequencies of 667 cm^{-1} and 2349 cm^{-1}. How would you assign these to the corresponding vibrational normal modes? Explain your answer. [5 *marks*]

(c) State the expression for the energy levels of the quantum harmonic oscillator. Describe the main features of this expression including the zero point energy, the spacing between adjacent levels and the number of levels. Consider the lowest frequency vibrational mode in CO_2. In a gas of CO_2 molecules in equilibrium at 1000 K, what is the ratio of CO_2 molecules in the first excited state to those in the ground state. [6 *marks*]

(d) The N_2O molecule is a linear molecule that has three infrared active modes at frequencies of 598 cm^{-1}, 1285 cm^{-1} and 2224 cm^{-1}. What is the structure of the molecule? Explain your answer. [5 *marks*]

(e) For N_2O, in addition to the three main infrared absorptions discussed in part (d), other peaks of weaker intensity appear in the infrared spectrum at 1167 cm^{-1} and 2563 cm^{-1}. Can you assign these? Explain your answer. [5 *marks*]

(f) N_2O is much more effective as a greenhouse gas than CO_2. Explain what a greenhouse gas is and suggest one possible explanation of this observation. [5 *marks*]

Structure

The paper has two sections, section A comprising a number of short questions, and section B, with a choice of one out of two questions. Although section A has six questions, each carrying five marks, the total (not *final* as it says in the paper) mark is capped at 20. This structure is an example of a compressor: if the majority of students achieve, say, 15 marks or more for showing what they know, they will have 30% of the marks on the paper and be well on the way to a pass. The choice element of section B does allow students to do well with only a limited knowledge of the curriculum.

Another structural comment is the time allowed; a significant issue with the section A mark compressor is that students can spend too long on that section. Pro-rata, allowing for 10 min reading time, students should spend 45 min on section A but there can be a tendency to spend longer, which means that there is less time available for the section B question.

Clarity of language

There are several cases of poor practice and inconsistency of nomenclature. In A5 we have 'the Planck's constant', an irritating grammatical error but also 'ϑ is the moment of inertia' without specifying the axis. In the same question, the masses given should be for the atoms, not the elements, and it is not clear what 'it' is referring to in the final sentence. A6 contains 'Assume that these two particles collide head-on with equal but slow, speeds.' The word 'slow' is meaningless without reference to something else—in this case, presumably, the speed of light.

In B2 (a) I would prefer not to refer to 'the associated classical speed of the object in its ground state', rather 'the classical speed of the object if it had an energy equivalent to the QM ground state.' Part (ii) is an unclear mix of classical and quantum and one has to think very carefully what is being asked. Presumably, the speed 3 cm s^{-1} is a classical speed but the question is very unclear. At the end of the same question we have 'Do the quantum mechanical results in (i) and (ii) agree with our everyday experiences? Explain your answer.' There is a great deal wrong here! First, the answer to (i) is classical, not quantum mechanical. Second, I have no idea what is being asked for in the comparison with 'everyday experience': who has everyday experience of a quantum particle in a box? And, third, 'explain your answer' has no meaning.

B2 (c) has 'Explain your answer'. Also 'Explain why the results for the probability distributions contradict what would be expected for a particle that moves in the box according to Newton's laws' could be much better worded. The spelling of 'dimensional' is incorrect in B2(d), which is not ambiguous but perhaps betrays that the checking was less than thorough. More seriously, I do not really understand what the question is asking for at all. Are students supposed to use the de Broglie relationship or simply write down the wavelengths in terms of L? And the second part—'Describe another physical system for which the same relation between a wavelength and a distance L holds.'—is incomprehensible.

The wording of B3(a) is untidy but at least unambiguous but 'explain your answer' occurs again in parts (b), (d) and (e), without adding anything: I suspect the setter really means *show your reasoning*. In B3(c) the dimensionality of the harmonic oscillator is not specified and the use of the word 'describe' is not ideal, since there follows a list which seems to give the answers. Note the missing question mark at the end. In (e) it would be better to be more specific in asking candidates to assign the lines, but note that the answer to part (e) depends on the answer to (d), so there is the possibility of an error carried forward.

Link to learning outcomes
Of the 30 marks available in section A, I estimate 20 are bookwork and the other 10 are from problems seen before. Given the nature of section A, as a mark compressor, we might expect a heavy emphasis on knowledge and simple applications. Question B2, has 14 marks for bookwork, 5 marks for known problems, 4 marks for unseen problems and 7 marks for explanations. Finally, B3 has 12 marks for bookwork and 18 marks for known problems.

In total, there are 46 marks available for bookwork, 33 for known problems, 4 for new problems and 7 for interpretations or explanations. That represents more than half the paper simply reproducing facts and 88% of it from material that the student has been given. Not only does rote learning allow a pass, a student can achieve an outstanding first-class mark simply by remembering things.

I have perhaps been overly harsh in analysing this paper but it does illustrate many of the traps to avoid when setting questions.

4.5.2 A general paper

In the previous chapter, I highlighted the importance of the various aspects of problem solving in defining the skill set of a physicist and expressed some scepticism whether such skills can be tested reliably in the format of a traditional examination. Nonetheless, several universities do set such general papers and the one following is one of the better ones of its type.

Techniques of Problem Solving Year 3
2 hours
Answer four questions from each section. You may attempt more questions, but only the best four answers from each section will be counted.
All questions are marked out of 10.

SECTION A

1. Estimate the number of electrons in the water contained in a full bath.
2. The potential energy of a uniform spherical charge of radius a and total charge Q depends on a, Q and ε_0. Use dimensional analysis to find the dependence on these three quantities.
3. A pipe of outer diameter 2.2 cm transporting water at a temperature of 50 °C passes through a concentric circular hole of diameter 7.6 cm in a wall. The thickness of the wall is 25 cm, and the wall temperature is 21 °C. The hole is filled with insulating foam of thermal conductivity 0.036 W m^{-1} K^{-1}. At what rate is heat transported through the foam? (You may assume that the thermal conductivity of both pipe and wall are much greater than that of the foam.)
4. An open cubic tank with internal dimensions $15 \times 15 \times 15$ cm^3 and made of 1.5 cm thick steel is filled to the brim with water at 18 °C. The temperature then rises by 12 °C. The coefficient of linear expansion of steel is 1.2×10^{-5} °C^{-1} and the coefficient of volume expansion of water is 2.4×10^{-4} °C^{-1}. Will the water overflow the rim of the tank? If so, state the volume of water which overflows; if not, by how much will the surface of the water be below the top of the tank?
5. A positron in flight annihilates with a stationary electron. Two photons are produced, one travelling at 30° to the original positron direction, and the other at 60°. What was the energy of the positron, in terms of the electron rest mass energy?
6. A uniform bar of length L and mass M is pivoted about one end so that it can rotate in a vertical plane. The bar is suspended by a light spring of spring constant k attached at a point aL ($a \leqslant 1$) from the point of rotation. In equilibrium the bar is horizontal. The bar is displaced by a small angle θ and then released. Show that the bar undergoes simple harmonic motion and derive an equation for the frequency of the oscillations.
   The moment of inertia of a uniform bar of length L and mass M about one end is $\frac{1}{3}ML^2$.

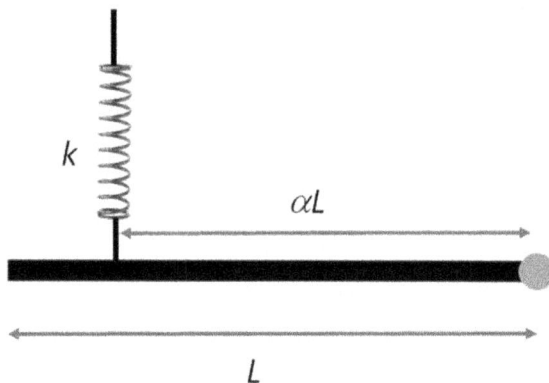

7. A container of height h is filled with a liquid. A small hole is formed in the side of the container at the level of the base. Show that liquid exits this hole with a velocity given by

$$v = 2gh.$$

You may assume the cross-sectional area of the hole is much smaller than that of the container.

The container now has a series of holes formed at different vertical heights above the base. Liquid exits these holes and falls onto a horizontal surface level with the base of the container. Use the above equation to find the height of the hole which results in the liquid hitting the horizontal surface the maximum distance from the wall of the container. For this condition at what angle and total speed does the liquid hit the surface?

8. The speed of sound in a gas can be written as

$$c = \sqrt{\frac{\gamma P}{\rho}}$$

where P is the pressure, γ is the ratio of specific heats and ρ is the density of the gas. A tube is open at only one end. It is filled with neon at 20 °C, and the fundamental resonant frequency is measured as 481 Hz. Subsequently, the same tube is filled with krypton at the same temperature. What is the new resonant frequency? How long is the tube? (You may ignore end effects.)

Neon and krypton are both monatomic ideal gases with molar masses 20.2 and 83.8 g mol^{-1}, respectively.

SECTION B

9. The Planck length l_P is the theoretical lower limit on measurable length. (A photon with a wavelength smaller than l_P would collapse into a black hole.) The Planck length is expected to depend on Planck's constant, $\hbar$, Newton's gravitational constant, G, and the speed of light, c. Use dimensional

analysis to determine the form of this relationship, and so make an estimate of the magnitude of l_P.

10. An experiment requires 180 litres of liquid hydrogen. The hydrogen is delivered in gas cylinders, each containing 50 litres under a pressure of 200 atmospheres at 16 °C. The density of liquid hydrogen is 70.8 kg m^{-3}. How many cylinders are required?

11. Spherical drops of mercury, each of radius a and carrying a charge Q, fall into a container with a square cross section of side length L. What is the vertical electric field after N drops have fallen into the container (assuming this is uniform across the container)? If ρ is the density of mercury and g the acceleration due to gravity, calculate the total vertical force acting on a drop as it falls, and hence find an expression for the maximum number of drops that will fall into the container.

12. A fractal can be created by taking a square (with sides of length 1) and adding four new squares to the mid-point of each side, where each new square has sides of length 1/3. Then squares of side length 1/9 are added to the 12 'exposed' sides of the new squares etc. The first three steps are shown in the diagram below. Derive equations for the area enclosed and perimeter after the Nth step. Hence show that whilst the area enclosed remains finite the perimeter tends to infinity in the limit $N \to \infty$.

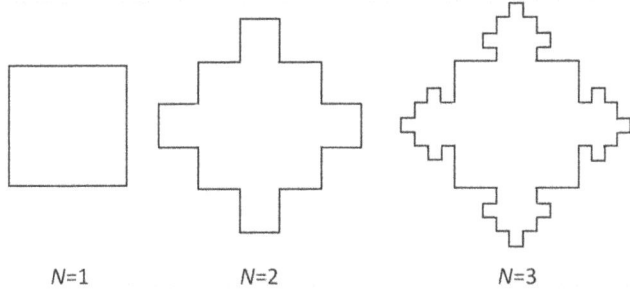

N=1   N=2   N=3

13. Estimate how long a line could be drawn with an entire pencil. The density of graphite (carbon) is 2.27×10^3 kg m^{-3}, and its atomic mass is 12. You may assume the minimum thickness of the line (perpendicular to the surface on which it is drawn) is one atomic layer. (For an order of magnitude estimate, you may treat the graphite as having a simple cubic crystal structure.)

14. A solid copper sphere, of radius 1.5 cm, is suspended by a thin insulating thread in a vacuum enclosure with walls close to absolute zero. If its initial temperature is 20 °C, how long will it take to cool to 0 °C? The density of copper is 8960 kg m^{-3}, and its specific heat capacity is 385 J kg^{-1} K^{-1}. State any assumptions or approximations you make.

15. When two lenses are initially placed 25 cm apart the final image from an object placed 10 cm to the left of the left-most lens is 30 cm to the right of the right-most lens. If the separation of the lenses is increased to 30 cm the

final image is now 20 cm to the right of the final lens. Given that both lenses are convex, find their focal lengths.
16. Upon heating, an element undergoes a phase transition where the crystal structure changes from face centred cubic to body centred cubic. If the density were to remain constant calculate the ratio of the sizes of the unit cells of the two structures.

Structure
There are 16 questions in total, split into two equal sections. Students are informed that they should answer four questions in each section. The differences between the sections are not immediately obvious; perhaps section B is more challenging.

With just 2 h for the paper and 16 questions, the reading time is around a quarter of an hour, leaving around 12 min per question, assuming a student tries only the required eight. In practice, students are likely to try more than that, further increasing the time pressure. Given that 'thinking like a physicist' takes time, the examination is a high-pressure affair with students possibly having to make the choice between persisting with a question after they become stuck and moving on to try another. It is not hard to imagine candidates becoming flustered and crashing badly.

The rubric states clearly that only the four best questions in each section will count, an arrangement that I think might exacerbate the pressure on the candidates. A different approach might be to create a mark compressor in which candidates are given credit up to 40 marks in each section but can accrue those marks from any of the questions. The advantage of this approach is that students can obtain credit for what they are able to do.

As a final remark, all questions are worth ten marks but are not of equal challenge. In particular, some, such as question 11, require the students to think carefully about what is going on, whereas with others, such as question 2, the approach is explicit in the question (students are also likely to know the answer.)

Clarity of language
Generally, the questions are clear about what is required. There were just a few ambiguities, most of a type where the candidate would probably make the correct assumption. In question 6 it would have been better to make it clear that the attachment of the bar allowed the bar to rotate freely about that point. In the second part of question 7 it should have been made clear that the height of fluid in the container is maintained constant and in question 11 the nature of the container—insulating or conducting—should be mentioned. A more minor point is that, these days, g is referred to as the gravitational field strength. In question 15 I assume the lenses are on the same axis.

Link to learning outcomes
The paper attempts to assess the students' ability to solve problems, make estimates, introduce sensible assumptions etc. The emphasis is correctly on the problem solving aspects rather than testing advanced physics. In other words, the students are

attempting questions where they should have a sound, mature grasp of the physics; they are not struggling with material they have only recently met.

The questions are well chosen in that, in most cases, they require candidates to identify what physics is involved, apply some equations and come up with a solution. Question 11 is a fine example: it requires students to realise that, since mercury is a conductor, the charge will accumulate at the surface and to use Gauss's theorem to determine the electric field. They will also need to make assumptions to neglect certain effects, for example to treat the mercury drops as point objects.

While the paper is an excellent example of its kind, I note that there is almost no test of synoptic thinking: each question relates to a fairly narrow area of physics. I suspect that the reason for that is to keep the knowledge required to a minimum to retain the emphasis on process rather than any sort of rote learning. In fact, because they are narrowly focused and do not rely on rote learning, they would make very good questions in the relevant module examinations.

There remains the question I raised earlier, which is whether a timed examination is the best means of assessing problem solving. With only 12 min available for each question, and that is assuming only eight are attempted, there is precious little time for thought. While one might say that one virtue of a timed examination is that it requires students to think quickly, the nature of problem solving does not lend itself well to this format.

As a final comment, in terms of driving student activity, a paper of this type only makes sense if it is used as an assessment of the skills acquired in training; that is, it should not be expected that students develop these tools simply by attending lectures. To reinforce my earlier point, assuming that students are being given classes in problem solving, some sort of continually assessed element within those classes is an essential adjunct to, or even replacement for, the timed examination.

4.5.3 Year 2 paper

Timed examinations are more suitable for some topics than others. One of those where it is possible to require less rote learning is quantum mechanics and the next paper is a second year one on that topic.

Quantum Mechanics Year 2
3 hours

Section A (50 marks)—Answer ALL questions in this section

[A1] (19 *marks*) A particle has the initial wave function $\Psi(x, 0) = Ae^{-k|x|}$, where k is a positive real number.

(i) (6 *marks*) State the meaning of the following terms used in quantum mechanics:
    (a) Wave function
    (b) Expectation value
    (c) Stationary state

(ii) (4 *marks*) Find the value A such that the wave function is normalized.

(iii) (6 *marks*) Find the expectation values of x and x^2. Using the uncertainty principle, find a lower bound on the expectation value of p^2, given that $\langle p \rangle = 0$.

(iv) (3 *marks*) Using a sketch compare the spatial part of the wave function at $t = 0$, with the wave function immediately after the position of the particle has been measured.

[A2] (14 *marks*) The spatial wave function of a quantum harmonic oscillator with natural frequency ω and mass m are:

$$\Psi_n(x) = \left(\frac{m\omega}{\pi\hbar}\right)^{1/4} \frac{1}{\sqrt{2^n n!}} H_n(\xi) e^{-\xi^2/2} \quad \text{where } \xi = \sqrt{\frac{m\omega}{\hbar}} x$$

and the corresponding energy levels are:

$$E_n = \left(n + \frac{1}{2}\right) \hbar\omega.$$

(i) (4 *marks*) Calculate the zero-point energy in suitable units, if the particle is an electron and $\omega = 1 \times 10^{15}$ rad s^{-1}.

(ii) (4 *marks*) Show that Ψ_0 and Ψ_1 are orthonormal, given $H_0(\xi) = 1$ and $H_1(\xi) = 2\xi$.

(iii) (6 *marks*) Show that the classically-predicted amplitude of such an oscillator with energy equal to the ground state energy is $|x_{\max}| = \sqrt{\hbar/m\omega}$, and calculate the probability of a ground-state quantum particle being found outside this limit.

[A3] (17 *marks*)

(i) (4 *marks*) Show that de Broglie waves have group velocity $v_g = \frac{d\omega}{dk}$ twice their phase velocity $v_p = \frac{\omega}{k}$.

(ii) (4 *marks*) Sketch the following 'step' potential and the real part of the spatial wave function solution for it, for particles of energy $E > V_0$ and $E < V_0$ approaching from the $-x$ direction. You may assume $V_0 > 0$.

$$V(x) = \begin{cases} 0; & x < 0 \\ V_0; & x \geq 0 \end{cases}$$

(iii) (5 *marks*) Show that the reflection coefficient in the case $E > V_0$ is given by

$$R = \left(\frac{1-\rho}{1+\rho}\right)^2 \quad \text{where } \rho = \sqrt{\frac{E - V_0}{E}}.$$

(4 *marks*) State without calculation the reflection coefficient when $E < V_0$, and explain your answer.

Section B (50 marks)—Answer ALL questions in this section

[B4] (16 *marks*) Consider a particle in the state

$$\Psi(x) = \sqrt{\frac{1}{7}}\Psi_1(x) - \sqrt{\frac{2}{7}}\Psi_2(x) - i\sqrt{\frac{3}{7}}\Psi_3(x) - e^{i\frac{\pi}{3}}\sqrt{\frac{1}{7}}\Psi_4(x)$$

where $\Psi_1(x)$, $\Psi_2(x)$, $\Psi_3(x)$ and $\Psi_4(x)$ are normalised energy eigenstates corresponding respectively to the eigenvalues $E_1 = E$, $E_2 = 4E$, $E_3 = 6E$ and $E_4 = 7E$.
  (i) (2 marks) What is the average energy $\langle \hat{H} \rangle$?
  (ii) (3 marks) What is the average squared energy $\langle \hat{H}^2 \rangle$?
  (iii) (3 marks) What is the spread of energy $\Delta \hat{H}$?
  (iv) (4 marks) What is the probability that a measurement of energy will find the energy smaller than E?
  (v) (4 marks) If at time $t = 0$ the particle was prepared in the state Ψ, what is the state at time t?

[B5] (10 marks) Let $\hat{p}_x$ and $\hat{p}_y$ be the momentum operators in the x and y directions, respectively.
  (i) (3 marks) Compute the commutator $[x^2, \hat{p}_x]$.
  (ii) (3 marks) Let $\hat{p}_\vartheta = \cos(\vartheta)\hat{p}_x + \sin(\vartheta)\hat{p}_y$ be the momentum operator in the x, y plane in the direction making the angle ϑ with the x axis. Compute the commutator $[\hat{x}, \hat{p}_\vartheta]$.
  (iii) (4 marks) Let $\hat{L}_1 = \cos(\vartheta)\hat{L}_x + \sin(\vartheta)\hat{L}_y$ be the angular momentum operator along an axis in the x, y plane making the angle ϑ with the x axis and $\hat{L}_2 = -\sin(\vartheta)\hat{L}_x + \cos(\vartheta)\hat{L}_y$ be the angular momentum operator along an axis in the x, y plane making the angle $\vartheta + \frac{\pi}{2}$ with the x axis. Compute the commutator $[\hat{L}_1, \hat{L}_2]$.

[B6] (17 marks) Consider a hydrogen atom prepared in the state

$$\Phi(r, \vartheta, \varphi) = \frac{1}{\sqrt{b^3 \pi}} e^{-\frac{r}{b}}$$

where b is some arbitrary distance not equal to the Bohr radius.
  (i) (3 marks) Show that the state Φ is an eigenstate of $\hat{L}_z$ the angular momentum around the z axis, and find the corresponding eigenvalue.
  (ii) (2 marks) What is the average value of $\hat{L}_z$ in the state Φ?
  (iii) (7 marks) What is the probability that a measurement of energy will find the particle with energy corresponding to the ground state? Recall that the ground state is

$$\Psi_{1,0,0}(r, \vartheta, \varphi) = \frac{1}{\sqrt{a^3 \pi}} e^{-\frac{r}{a}}$$

where a is the Bohr radius.
  (iv) (2 marks) Let $\Psi_{n,l,m}$ be the eigenstates of the hydrogen atom. Give an example of an eigenstate which is orthogonal to the state Ψ and give a short explanation of your choice.
  (v) (3 marks) Helium is an atom whose nucleus has two protons, hence the neutral helium atom contains two electrons as well. We want to completely ionize the helium, by first removing one electron then the second. Which of these electrons costs more energy to remove, the first or the second? Give a short explanation of your reasoning.

[B7] (7 marks) Consider two particles of masses $m_1 < m_2$ in a 1-dimensional box of length L situated between $0 \leq x \leq L$. The particles are non-interacting with each other. Write the first excited state of this system and its energy.

Hint: Recall that for a single particle of mass m in this box the eigenstates are

$$\Psi_n(x) = \sqrt{\frac{2}{L}} \sin\left(\frac{n\pi x}{L}\right)$$

and the corresponding energy eigenvalues are

$$E_n = n^2 E_1 \quad \text{with} \quad E_1 = \frac{\hbar^2 \pi^2}{2mL^2}.$$

Structure
The paper is unusual in not offering candidates any choice at all. That is a perfectly defendable position in terms of encouraging students not to omit any part of the syllabus but it has the disadvantage of not allowing the weaker students to accumulate some marks using a sort of mark compressor as seen in some other papers.

Another unusual feature is that there is no common tariff to the questions. That does not lead to any issues of balance, due to the absence of choice, but it does mean that the candidates will find it harder to plan their time. However, the three hour length is reasonably comfortable for this number of questions.

One puzzling element of the structure is why there are two sections. Both have equal weight in terms of marks and all questions in each section are compulsory. Perhaps the structure is because different people taught the material for the two sections but I am not able to find any obvious reason for the separation in terms of the examination.

Clarity of language
In general, the questions have little ambiguity, which illustrates that it is much easier to be clear in defining mathematical questions than those that require physical understanding and interpretation. Most of the questions across the paper require mathematical manipulations and little else. While that is not a criticism of the paper, there are more qualitative questions one can ask about quantum mechanics; for example, the last part of B6 does require a qualitative explanation.

Question B6 (iv) does have some ambiguity. First, it is not quite clear what the wave function Ψ refers to; presumably $\Psi_{1,0,0}$ of the previous part. And 'a short explanation of your choice' might be better as 'justify your choice'.

Link to learning outcomes
The essence of a mathematical paper is that it is relatively easy to set unambiguous questions and, for this paper, the best strategy for the student is to practice with set problems, which is no bad thing. However, there is little opportunity for students to display intuitive interpretations.

What is perhaps lacking in the form of assessment are any computer-aided solutions, an absence that refers to the form of the assessment rather than being an inadequacy of the paper.

4.5.4 Option paper

As a final example, I reproduce an examination for an advanced option.

Imaging and Manipulation at the Nanoscale: Year 3 Option
90 minutes
Answer 3 out of 5

1.
   (a) A metallic tip is held above a conducting surface such that the tip–surface separation, z, is less than 1 nm. A voltage V is applied across the tunnel gap leading to the flow of a current, I. Explain why an exponential dependence of the current on tip–sample separation would be expected if the current flow is due to quantum mechanical tunnelling across the tip–surface junction. {3 *marks*} How is the exponential decay factor related to the work function, W? {2 *marks*}

   (b) Using simple arguments explain how the dependence of I on z can be used to estimate the work function, W, and show that {5 *marks*}

   $$W = \frac{\hbar^2}{8m_e}\left(\frac{\mathrm{d}\log_e I}{\mathrm{d}z}\right)^2.$$

   (c) Using a piezoelectric transducer the tip is advanced towards the surface through a displacement of 0.4 nm. The current measured as a function of position is shown in the figure. Use this plot to estimate the value of the exponential decay factor and the work function (you may assume that the tip and the sample are formed from the same metal. {8 *marks*}

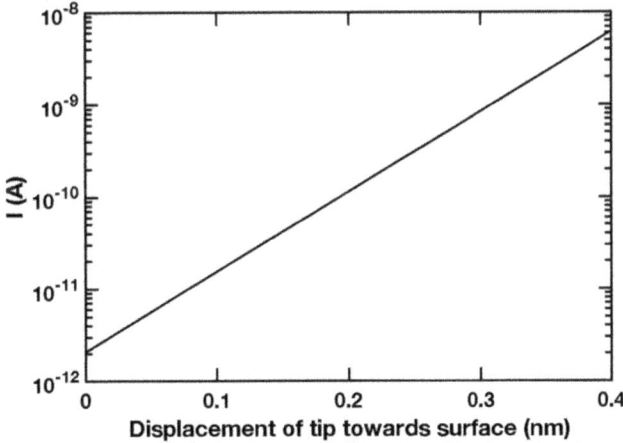

(d) Explain what is meant by the constant current mode of scanning tunnelling microscopy (STM) including in your answer a brief discussion of feedback, the target tunnel current (setpoint) and the interpretation of image contrast in this mode. {4 *marks*} The tip is scanned parallel to the surface and images are acquired in the constant current STM mode. The setpoint is changed from 0.1 to 1.0 nA. From the plot in the figure state whether the tip is displaced towards or away from the surface and estimate the magnitude of this displacement. {3 *marks*}

2.
(a) Explain what is meant by an unreconstructed surface {1 *mark*} and sketch the arrangement of lattice points on the unreconstructed (100) and (111) surfaces of a face-centred cubic crystal. {2 *marks*} What is a reconstructed surface? {1 *mark*} Give one example of a metal surface which is unreconstructed and one example which undergoes a reconstruction. {2 *marks*}

(b) The Si(111) surface can undergo a (7 × 7) reconstruction under the appropriate preparation conditions. Sketch the arrangement of the Si adatoms which form the top layer of this reconstructed surface and explain briefly the bonding arrangement of the Si adatoms and how this leads to a half-filled 'dangling' bond oriented normal to the surface plane. {3 *marks*}

(c) In the Tersoff–Hamann model of the tip–sample junction of a scanning tunnelling microscope (STM) the current flowing between the tip and surface is proportional to the local density of states, $n(r_0, E_F)$, where E_F is the Fermi energy and r_0 is the centre of the spherically symmetric potential which models the apex of the tip. Explain how $n(r_0, E_F)$ is related to the quantum states of the sample surface. {3 *marks*}

(d) The Tersoff–Hamann model was originally developed within the assumption that the applied voltage between tip and sample is small. Explain how this model may be extended to imaging when the tip–sample voltage is not small. {2 *marks*} Explain why images, for example of semiconductors, can appear different depending on whether a positive or negative voltage is applied to the sample (with the tip held at zero voltage). {2 *marks*}

(e) Explain why the adatoms of the Si(111)–7×7 reconstruction may be resolved in constant-current STM images using either positive or negative sample polarity. {2 *marks*} In a semiconductor sample which is doped n-type through the addition of antimony (Sb) impurities, some of the Si adatom sites appear as low contrast features in images acquired using a positive sample voltage. Explain why this occurs (you may assume that Sb atoms substitutionally replace Si adatoms while retaining the same bonding arrangement; note that Sb is a Group V element and therefore has one extra valence electron as compared with silicon).

{3 *marks*} Would you expect these features to appear bright or dark when imaged with a negative sample bias? Include your reasoning in your answer. {2 *marks*}

(f) With the STM tip held at a fixed height above the Sb-doped silicon the current, I, is measured as a function of sample voltage V over a range from -4 V to $+4$ V. Sketch the expected $I(V)$ curve. {2 *marks*}

3.

(a) Scanning tunnelling spectroscopy (STS) is a variant of scanning tunnelling microscopy (STM) which can be used to measure the energies of the quantum states of an adsorbed molecule on a surface. This is achieved by holding the tip of the STM at a fixed height above the molecule and measuring the current, I, flowing between the tip and surface as the voltage, V, applied to the sample is varied (the tip is earthed). Explain how the differential conductance, dI/dV, can be related to the energies of the states of the molecules. Include in your answer relevant band diagrams at different voltages. {6 *marks*}

(b) A tip is held above a planar molecule and STS measurements show peaks in dI/dV for sample voltages of 1.3 V and -0.5 V. Determine the energies of the highest occupied molecular orbital (HOMO) and lowest unoccupied molecular orbital (LUMO) relative to the Fermi energy of the substrate and state the HOMO–LUMO energy gap. {2 *marks*} What scanning voltages should be used in order to acquire images of the probability densities of the HOMO and LUMO states? Explain your reasoning. {2 *marks*}

(c) After heating, several molecules fuse together to form a rectangular metallic strip with lateral dimensions $L_x \times L_y$. Assuming that the edges of the strip can be modelled as infinite potential barriers show that the electrons confined in this metallic region are quantised in states with energies

$$E_{kl} = E_0 + \frac{k^2 h^2}{8m^* L_x^2} + \frac{l^2 h^2}{8m^* L_y^2}$$

where l, k are integers ($l, k > 0$) and m^* is the effective mass of the electrons and E_0 is a constant. {3 *marks*}

(d) An STS measurement is performed on two strips with different dimensions. The lowest peak in dI/dV occurs at -0.050 V for a strip with dimensions $L_x = 5$ nm and $L_y = 10$ nm and at $+0.010$ V for a strip with $L_x = 3$ nm and $L_y = 6$ nm. Calculate the constants E_o and m^*. {7 *marks*}

(e) On another part of the surface many molecules fuse together making a semi-infinite metallic region bounded by a straight edge. A set of standing waves running parallel to the edge are observed when imaging in constant current mode STM with a small bias (so that states close to the Fermi energy dominate the current flow).

Explain the physical origin of these standing waves {2 marks}; calculate the expected wavelength of the standing waves. {3 marks}

4.
(a) Describe the operation of an atomic force microscope (AFM) which can be used to acquire images of surfaces in contact mode (you may assume that the AFM is operated under ambient atmospheric conditions at room temperature and that micromachined cantilevers are used as force sensors). Include in your answer a diagram in which the important components of the AFM are labelled. {6 marks} Sketch the form of the typical dependence of the tip–surface force on their separation and identify on your diagram the regime where contact mode AFM is performed. {2 marks}

(b) Assuming that the repulsive potential energy between a surface and a cantilever can be modelled using the repulsive part of the Lennard-Jones potential,

$$V(z) = E_0 \frac{z_0^{12}}{z^{12}}.$$

Calculate the tip–surface force as a function of z, the tip–sample separation (E_o and z_o are constants with units of, respectively, energy and length). {1 mark}

(c) The cantilever is scanned across a boundary between two different materials. The value of z_o is the same for both materials but the value of E_o is different and is denoted, respectively, E_{oA} and E_{oB} for materials A and B. Derive an expression for Δz, the topographic height difference across the boundary between the two materials when this region is imaged using contact mode AFM and a target force (setpoint) F_o. {5 marks}

(d) Calculate the step height D_z if $E_{oA} = 2$ eV, $E_{oB} = 1$ eV, $z_o = 1$ nm and $F_o = 100$ pN. {4 marks} If the spring constant of the cantilever $k = 1$ N m^{-1} what is the cantilever deflection? {1 mark}

(e) Show, explaining your physical reasoning, that the root mean squared amplitude of fluctuations of a cantilever due to Brownian motion is given by

$$z_{\mathrm{rms}} = \sqrt{k_B T / k}$$

where T is the temperature and k_B is the Boltzmann constant. {4 marks}

(f) Calculate the temperature for which $z_{\mathrm{rms}} = D_z$. {2 marks}

5.
(a) A quartz tuning fork with an attached tip may be used as a force sensor in atomic force microscopy. Describe briefly how they are used {2 marks} and give one advantage of quartz tuning forks over the other commonly used sensors, silicon cantilevers. {1 mark}

(b) Quartz tuning forks may be used since their resonant frequency is shifted in the presence of a spatially varying force $F(z)$ (z is the separation of the tuning fork sensor and a surface of interest). Show that the shift in resonant frequency, $\Delta\omega_o(z)$, due to the interaction between the tuning fork and the sample surface is given by

$$\Delta\omega_o(z) = -\frac{1}{2}\frac{\omega_o}{k}\frac{dF}{dz}$$

where ω_o is the free resonant frequency (i.e. the resonant frequency in the absence of the sample) and k is the spring constant of the tuning fork (you may assume that the amplitude of oscillation is small). {8 marks}

(c) A single electron is trapped at the sample surface at a position with Cartesian co-ordinates (0, 0, 0). Through a contact electrification process it is possible to add an electron to the end of the tip attached to the tuning fork sensor. Derive an expression for the frequency shift of the tuning fork sensor as a function of z as it is advanced and retracted directly above the trapped charge on the surface (i.e. above a lateral position with x, y co-ordinates (0, 0)). You may assume that screening of charges from the substrate and tip may be neglected so that the interaction can be modelled as a Coulomb interaction between two charges in free space. {6 marks}

(d) A tuning fork with a free resonant angular frequency of 6×10^4 radians s^{-1} is positioned 2 nm above the charge and a frequency shift of 1.7 radians s^{-1} is measured. Calculate the spring constant of the tuning fork. {3 marks}

(e) Through a surface modification process a second electron is positioned on the surface separated from the original charge by a distance d and with co-ordinates (d, 0, 0). The tip is placed midway between the two charges and then retracted perpendicular to the surface. Derive an expression for the frequency shift of the tuning fork sensor as a function of z along this trajectory. {5 marks} [Hint: only the z-component of the force contributes to the frequency shift.]

Structure
As is relatively common in more advanced papers, there is only one section but with a choice of three out of five questions, each carrying the same number of marks. The questions are well-constructed, each dealing with a single topic, in most cases with the separate sections following on from one another. Although the absence of a section A/B structure does not allow students to accumulate marks by showing what they know across a range of topics, each question has straightforward components to allow all candidates to collect credit. Also, with a generous choice allowing two topics to be ignored, some students are likely to omit some material from their revision schedule.

The paper is very long—five full A4 pages—for a 90 min session. The rubric suggests spending 30 min on each question but there is a considerable reading load, possibly requiring 15 min of deliberation. In a high stakes paper, candidates need to be sure that they have the confidence to attempt a question so they would be well-advised to read the paper thoroughly before beginning. In addition, I estimate that even the physical challenge of writing the 15 pages or so of answers would be hard to accomplish for some students. It would make sense to allow some reading time in advance of the candidates being allowed to start writing. Note also that, for students requiring extra time to complete the assessment (for example those with disabilities), the extra time is usually proportional to the duration of the paper, not its word length. For someone who is, say, dyslexic, even with more time, the length of this paper would be an issue.

Clarity of language
Part of the reason that the paper is so long is that the setter has taken care to be as specific as possible in defining the questions as unambiguously as possibly; there are very few places where the language is unclear. In addition, there are several hints and a couple of wordy introductions that contribute to the length of the paper.

I found only one minor typo: a bracket missing in Q1, although there were a few places where extra punctuation would have helped the clarity of what was being asked.

Slightly more seriously, in two places (Q2d and Q5b), some parameter is referred to as small without reference to anything else. In 2d, an applied voltage is said to be small and in 5b it is the amplitude of oscillation of the tuning fork. In the latter case, it might be assumed that the amplitude should be small compared with the separation of the tuning fork and the surface but in the former, the comparator is much less obvious. In practice, these omissions are unlikely to lead candidates astray, because the assumptions of smallness would have been made in the lecture notes, but that statement then hints towards the exercise being one of rote reproduction.

In a couple of places, it is not clear that a section is using information from a previous one. In 3e, the value of E_o has to be used from the previous section. In 5d and 5e candidates require the formula derived in 5c—the formula is not given so there is the potential for candidates to have consequential errors.

Link to learning outcomes
There are some very good features of this paper. Primarily, it does contain some synoptic elements, drawing upon the student's knowledge of quantum and classical mechanics as well as electromagnetism, albeit at elementary levels, respectively: particle in a box; force as the derivative of potential energy; and Coulomb's law. In addition, there are examples involving numbers in realistic situations, although many of the numerical tasks require little more than the substitution of numbers into given equations; there is one question, 5e, that requires taking the derivative of a function using the chain rule as well as some conceptual grasp of the problem to be solved.

Perhaps less positively, the paper is bookwork heavy and a candidate could achieve a first-class mark by rote learning coupled with the ability to use a calculator.

Together, with the generous choice that allows candidates to ignore two themes, the best approach in preparation for the test would be to question-spot and rote-learn material for the topics that are likely to appear.

These comments are not meant to imply that this is a bad paper; indeed, it has been set with care with very little ambiguity and includes interesting, well-constructed questions, a fine exemplar of its type. Rather, the commentary indicates the limitations of the timed examination format in promoting desirable student outcomes. It is an ironic feature of examination assessments in the UK that the more advanced a paper, particularly when it refers to an optional module, the more it depends on rote learning for candidates to perform well. The problem solving skills that we all value so much in physicists are seldom required for the very simple reason that the format does not allow it.

4.6 General remarks

The inclusion of such an extensive discussion on timed examinations reflects the general weight given to timed examinations within the university system, rather than any recommendation of them as the best method of assessment. Anyone in a physics department is likely to be asked to set written papers so one of my purposes was to determine what we mean by a good examination paper. I have tried to provide guidance and illustrated the points with examples of good and bad practice. Specifically, in technical terms, a 'good' paper should be:

- Free of error.
- Free of ambiguity.
- Specific on what is required from the student.
- Doable within the time available.
- Matched to the level of the students.
- Doable by someone who has worked independently.

In practice, in most UK universities the final condition is seldom achieved, mainly because the ubiquitous bookwork elements of the questions would be exceedingly difficult to answer for someone who had not been working from the official notes.

The penultimate condition is one that is universally applied but not always explicitly recognised. One of the most common issues when a new member of staff starts teaching and setting examinations is that they do not pitch the level correctly. Naturally, the reason for that is that they usually have experience only of the rarefied research environment and are not used to working with a broader range of ability. But this condition does not sit happily within an allegedly criterion-based assessment system.

The requirement of the second bullet, which, as we have seen in the examples, is not always satisfied, is ironic in the context of physics. Almost all physics problems one can think of are fuzzy to some extent, requiring for their solution synoptic thinking, approximations of one sort or another and, frequently, a compromise between tractability and exactitude. Yet the format of traditional examinations, set in fixed time slots, requires students to be absolutely clear what is required of them.

These examinations, for all their virtues of being easy to set and mark and providing a range of student marks are, in the end, a wholly artificial representation of the subject. The skills that enable a student to achieve a high mark are not necessarily those that lead to them being a successful physicist.

This brings me neatly to the second purpose of my analysis, which was to investigate what a traditional examination actually assesses. To a varying degree with all the examples above, there are three elements to the assessment:

1. Knowledge—how much the candidate knows. In most cases these days, formulae are provided so the knowledge under test is related to physical laws, techniques, concepts etc. Generally, this type of question does not explicitly require word-for-word reproduction of the notes, although that is not precluded, more an explanation in the candidate's own words.
2. Reproducing derivations of formulae. This common type of question encourages short-term rote learning and probably discourages broader reading around the subject, although I accept that one can argue that a good understanding of the derivation makes remembering it considerably easier.
3. Solution of simple problems, often either substitution of numbers into equations or solving a problem similar or identical to one that the candidate has seen before.

That is more or less it. I should say that these are not negligible virtues to examine; it is important that physics graduates know stuff, that they know how important formulae are derived, and that they are numerate and have some ability to perform simple manipulations. However, I think I would argue that those three things are not so important that they should provide such a large fraction of the assessment portfolio as they do in most physics departments.

The main constraints of this form of assessment are the fixed time, which precludes deep thinking, and the need to maintain isolation without the aid of electronic devices, which precludes the setting of computer-based tasks and the possibility of discussing problems with other students. One can argue that written communication skills are also part of the assessment but my experience is that few if any markers penalise poorly written or ungrammatical answers; generally they are looking to award marks, not take them away.

Even within those constraints, it is possible to see ways of improving the papers. One is to increase the number of questions that test conceptual understanding. This type of question, exemplified by the various concept inventories (a recent review can be found in [1]), generally presents a qualitative problem in which candidates are asked to say what will happen in a given situation but without a need for a numerical solution. Eric Mazur has an example he uses on expansion, where he says that there is a sheet of metal with a hole in it which is warmed up: what happens to the diameter of the hole? (See [2] for a typical example of this question.) That is an elementary level question but it is perfectly possibly to generate similar types of question for more advanced topics. For example, in quantum mechanics one could ask candidates to sketch the wave function of an electron in a finite, 1D potential well and then ask how they would expect it to change if a positive charge of diameter one fifth of the width of the well is introduced in the centre of the well. In both cases, within a formal examination, it is only possible to answer the question intuitively. Very few of the papers reproduced above contained this type of question.

The second way in which the examinations could be improved is for them to contain more synoptic material. The great barrier here in the UK is the modular system, where it is accepted that the assessment of a module can only relate to the material taught within it. Of course, physics is a cumulative subject and it is the case that basic physical ideas can be assumed in more advanced modules but it is rare indeed for any synoptic assessment of material from two modules at the same level. I am no particular fan of the modular system and it is possible to subvert it by blocking together modules: if the basic unit is 10 credits (in a model where 120 credits constitutes a full year) 30 or 40 credit modules on general topics such as, say, classical physics, could provide an environment for a more synoptic approach.

One of the technical virtues of the timed examination is that it ensures that the work produced is the candidate's own. However, the recent SARS-CoV-2 pandemic has had a radical effect on the spring 2020 examinations with many universities switching to open-book arrangements. The short notice of that change meant that many of the examinations were not properly thought through as open-book and still contained varying degrees of bookwork. However, the ongoing nature of the pandemic has meant a similar approach was applied also to 2021, when people had more time to think about how to assess.

In fact, the difference between a closed and open examination turns out to be much less than people expected. It is quite difficult and time-consuming to find reliable solutions on the Internet and even sifting through notes takes time if you are not thoroughly aware of the material. The main issue at my own university turned out to be student collusion, either by texting as part of a group or even with students sitting and working together. As a result, some universities are looking very carefully at their assessment regimes with perhaps a move away from the traditional approach.

What is clear is that there are many desirable qualities that we would hope graduates to possess that are not being assessed through this route. These include all aspects of experimental work and investigative projects as well as the range of generic skills, including group work, communications and computing. These form the topic of the next chapter.

Bibliography

[1] Laverty J T and Caballero M D 2018 Analysis of the most common concept inventories in physics: what are we assessing? *Phys. Rev. Phys. Educ. Res.* **14** 010123

[2] Harcy V From passive listener to active learner: using peer instruction to increase student engagement *CVM Teaching Academy*

IOP Publishing

Assessment in University Physics Education

Peter Main

Chapter 5

Experimental and other skills

5.1 Investigative work

Experiment is at the heart of physics and is an essential part of any programme. Even those pursuing a theoretical physics degree must have some appreciation of the nature of, and the limitations of, experimental work. However, I think the label of 'experimental' or 'practical' has led to this type of work being seen as somehow separate from the 'proper' physics covered in the theory parts of the course. The reality, of course, is that the two are inextricably linked and the interplay between theory and experiment is a defining feature of the subject. For that reason, and because I think it is more appropriate term, I prefer the term 'investigative' to describe the whole process by which the subject progresses. Whether an experiment throws up a result that leads to a theoretical conundrum which in turn produces a major advance, as in the discovery of the neutrino, or an experiment confirms a theoretical prediction, as with the discovery of gravitational waves, theory and experiment go together and that should be reflected in the way we teach and assess investigative activity.

Within a physics programme, investigative work provides an opportunity to develop some of the desirable features of a physicist that are not always easy to incorporate in exam-assessed theory modules. Some of particular significance are: to carry out work of a synoptic character, bringing together different topics; the ability to make estimates and perform back-of-the-envelope calculations; and to integrate computing and communication skills in a natural fashion, i.e. without having to teach those elements without a context.

Finally, the development of investigative skills is a cumulative process, starting with an awareness of what it is to do measurement in physics and culminating in an independent investigation either individually or in a pair or larger group. One should not expect students to make this journey on their own; there needs to be a coherent programme of instruction and, as ever, the assessment is crucial for determining student activity.

As with any type of assessment, a good starting point is to think about what the purposes of the activity we are trying to assess are. I suggest the following list is reasonably comprehensive:

- *Illustrating physics.* For many students, me included, actually doing or seeing something for themselves is highly beneficial to their appreciation of the subject. Even such simple experiences as feeling magnetic repulsion or observing optical fringes offer much more than any explanation or picture can provide. Despite the undeniable importance of this purpose, there is no obvious need to assess it. I am aware of some departments that include the marks from practical work as part of the overall grade for a theory module. That is, an experiment on, say, Newton's rings would be assessed in some way and those marks would then contribute, say, 30% of the total for a module on optics. While it is no bad thing to have a variety of assessment within a module, I am not enamoured of this approach. I think it splinters the coherence of the investigative approach and the development of skills. Although the illustration of physics is an essential product of doing experiments, it is not why we do them as professional physicists.
- *Familiarisation with apparatus.* It is essential that students have some experience of, and are able to use reliably, the types of apparatus used to make measurements of various parameters. One can argue that these days experiments are all automated and there is no need for anyone to press buttons or twiddle knobs on voltmeters or frequency meters. My view is that the computerisation of laboratories makes it even more desirable that students are aware of what is happening at the front end of the measurement. While some of this outcome will occur naturally as students use the apparatus, in some cases one can imagine a threshold competence test, of a pass/fail variety for particularly important types of kit.
- *Handling data.* Measurements yield numbers, which have to be processed. There are a number of important steps: first and foremost, the data have to be reliable and valid, so students must learn how to check for reproducibility and, crucially, be able to estimate the uncertainty in the numbers they produce. The latter is often referred to as error analysis, a term to be avoided as no mistakes are involved. They should be able to manipulate their data to compare with whatever theory is available and produce a suitable comparison. The use of computational techniques and dedicated software should be standard.
- *Experimental design.* At the earliest stage, experimental design may be as simple as deciding which parameters to keep constant, which to vary and which to measure. For example, if one wishes to measure the I–V characteristics of a device as a function of magnetic field and temperature, it is important to realise that the system takes time to reach thermal equilibrium so the temperature should be fixed; the other parameters are much easier to vary. In addition, students should be aware that a measurement may perturb a system. Continuing with the previous example, changing the magnetic field might induce eddy currents in the apparatus, which then affect the temperature.

Slightly more advanced tasks might include the ability to select a set of apparatus from a pool to test a certain hypothesis and, at the most advanced level, designing an investigation from scratch.

- *Scientific methodology.* Or, in a nutshell, how science works. This topic has a whole branch of philosophy dedicated to it but it is fair to say that the philosophical discussion has had little effect on what scientists actually do. In the past, students' appreciation of the scientific method has been hindered by the habit of asking them to do set experiments; such an approach can often help them understand uncertainties, handle data etc, but there is always a 'right' answer at the end of it. In real science, we do not do experiments to determine something we already know. At the early stages of learning we cannot expect students to do real research, but we can introduce investigative elements into the programme so that something of the interplay between theory and experiment is revealed—a good exercise might be to test an incorrect theory and it is easy to think of simple experiments to do just that.

- *Scientific integrity.* Another consequence of the traditional approach of set experiments with their expected results is there is no incentive for anyone to maintain scientific integrity: so long as they obtain the right answer, what does it matter if they throw away a result or two or if they massage a data point on a graph to make it fit the line? Teaching integrity is not easy [1], but there are ways that assessment can encourage it. There are many examples in history of scientific fraud motivated by career pressure, political interference or simply because an investigator cannot bear their theory to be incorrect [2].

- *Keeping a record.* Central to maintaining scientific integrity is the need to keep an accurate and comprehensive record of the experiment. There can be several elements to this, including the conditions under which measurements were made, sample number, time, values of fixed parameters etc. It is also good practice to analyse results, for example by plotting graphs, as far as possible as the experiment is being performed. The traditional method of record-keeping is the bound laboratory notebook, which still has a number of advantages in the computer age: for example, one can easily inspect it at any time, one can see if changes have been made and everything is in one place. One can, of course, devise computer-based alternatives which have similar virtues but my own view is that, certainly at the first stages of learning, a paper notebook is the simplest and most effective method.

- *Imagination, synopsis and intuition.* Investigative work, by its nature, requires a synoptic approach. The real world does not package itself into neat, module-friendly topics. Tasks and assessments should be designed to reflect this, bringing in physics from different areas. Imagination and intuition are notoriously difficult to teach but they can be encouraged. Key is to ensure that, from the beginning of the programme, students are expected to think about the investigations rather than following recipes.

- *Communicating results.* The communication of results is an essential part of training. It includes the ability to provide concise descriptions and explanations and to display data in a visually attractive and informative manner.

In the UK there has been a tradition of so-called formal accounts in which students write about an experiment they have done; these accounts are seldom related to any form of real scientific communication, are universally hated by students, and are tedious and difficult to assess. More imaginative exercises can include podcasts, a paper in the format of a journal, web pages with 3D representations of data and so on. That is not to diminish the requirement for clear and logical descriptions but the variety of formats better develops skills appropriate for physics and more general applications.

The list above represents the learning outcomes for investigative work, although the first bullet, the illustration of physics, does not require assessment. Another point I would add to a successful course is that the students have the opportunity to discuss their work with each other, to talk about the physics, improve their understanding and learn from their colleagues. This remark applies as much to project work as it does to initial training and I shall return to the point later. For the moment, let me just say that the argument that a student must work independently for them to be assessed individually is flawed in many ways. Further, such an arrangement creates an artificial environment; no scientist makes progress without interacting with other people.

Another general point about laboratory/investigative work is that it inevitably requires some sort of demonstrator, that is a person who is available to give advice to students. Given the high intensity of such work, it is common to use paid postgraduate students to carry it out. In a traditional regime of classic, set experiments, this works well because the demonstrators can familiarise themselves with the details of the experiments. However, given that the traditional approach relates to few of the learning outcomes, it is sensible to move to a more open-ended approach to investigation, and that requires much more of the demonstrators, who must be carefully trained in the spirit of the approach. An anecdote helps: about 35 years ago, I replaced some of the set experiments in our Year 1 laboratory with open-ended, mini-projects, which required the students to test a theory using a given set of apparatus but without detailed instructions. The shift was, initially, not very successful as the postgraduate demonstrators quickly decided what they thought (incorrectly in most cases) was the best way to do the experiment and then effectively acted as an oral instruction manual for the students. I had ensured that the demonstrators were thoroughly conversant with the mini-projects but I had underestimated the need to train them how to coax decisions from students instead of just telling them the answers.

During the worst days of the Covid pandemic, many universities struggled to provide a hands-on laboratory experience. As a result, the sessions were postponed or, more commonly, replaced by basic, at-home experiments, online simulations or a combination of both. While the response to the Covid restrictions has led to a whole new approach to teaching theoretical material, an approach that I hope will continue when those restrictions are lifted, my own view is that there is no adequate replacement for a well-designed laboratory experience (the Open University, with its need to serve remote students has developed a high quality, online laboratory [3]). I hope that universities, suffering the financial effects of the lockdown, do not see online resources as a cost-saving substitute for provided dedicated laboratories, as appears to have happened in some schools, given the number of companies advertising online experiments at that level (see, for example, the Virtual Physical Laboratory [4]).

5.2 Projects

Physics programmes generally include a major investigative project, usually in the final year. In many countries, this involves an extended placement in a research group, with the assessment in the form of a dissertation. In the UK and some other countries, the nature of the project can vary a great deal from institution to institution. In this section, I consider what a project should deliver in terms of matching to the learning outcomes described above. Examples of assessments will be considered in a later section.

I use the word project to mean an investigation of a topic previously unknown to the students, through experiment, theory or a combination. That allows a broad range of activities, ranging from theoretical modelling to more applied tasks such as, say, constructing a device to perform a particular task. An essential feature is that students themselves have the knowledge and competence to make decisions about the direction of the project. They must also be allowed sufficient independence to exercise their choices.

It follows that there is no requirement for a final-year project to be leading-edge research, although there is nothing to preclude the topic of a project being relevant to a research group. However, there are undoubtedly dangers to this approach. I am aware of academics who insist that a final-year project must be undertaken by a student alone and must be based in a research laboratory. I can see the positive features: the student is exposed to the latest developments and may treat the exercise as a dry run for a possible PhD. In terms of assessment, by working alone, a student cannot ride on the back of a more able partner. However, I believe there are also substantial negative features to this arrangement.

First and foremost, the students are robbed of the opportunity of discussing the physics with another person who has the same level of knowledge and experience. If we are to satisfy my requirement that the project should allow the students to make decisions, that is a major barrier. Of course, when a student is based is a research group, there will be postgraduate students and post-docs who will be able to share their knowledge and expertise with students. Indeed, for some academics, this is another advantage, because it means they can delegate some of the advisory responsibility to these people, however dubious that might be in terms of quality management. But the drawback here is that the project student is not at the same level as the postgraduate students or the post-docs, so the conversations take on a completely different character than they would if two students argued the case between themselves. The researchers will be seen as sources of truth in a way that a fellow student will not.

Another disadvantage of the single-student placement in a research group is that it makes the project much more difficult to assess. The reason is that one can never be sure how much the student has been helped. No project sponsor wants their project to fail, so if a student is struggling it is natural (and commonplace) to ask one of the group to provide some direct assistance, inevitably producing a better outcome than if the student had been left to their own devices. It is very difficult to take such direct support into account, particularly if the main part of the assessment, as I shall recommend below, is carried out by someone not directly related to the project.

None of these objections suggests that it is unacceptable to have projects associated with research groups. Rather, they point to the necessity of ensuring there is a common set of requirements that all eligible projects should satisfy, including the essential point about students having the competence to make decisions and the consequent need to have a partner with which to discuss these decisions. Other members of the advisor's research group can play a role but it should be restricted to explaining how a piece of apparatus or program etc, works and not making a direct contribution to the project itself.

The above comments do not really apply to programmes that involve an extended placement in a research group, during which time the student does nothing in parallel and effectively becomes a member of the group. In that case, the student is effectively pursuing a research qualification, which goes beyond the remit of this book. Nor do the comments apply to external placements, in industry for example, which serve a different purpose, although they might apply to an externally sponsored project which actually takes place within the university.

5.2.1 Types of projects

Some university academics find it difficult to think of projects, particularly those that are outside their own research area, leading to some of the issues outlined above. In addition, in a training programme which is based around investigation, there is a need for projects at all levels and of different durations. The Internet has loads of examples of possible projects (see for example [5]) but I thought that it might be useful to give a few illustrations. The examples below are all ones I have used.

(a) Students are presented with apparatus and a theoretical equation, which they are asked to test. Almost any set experiment can be modified for this purpose. An example is to use a sonometer to determine the speed of sound along a wire as a function of the tension in the wire. All the apparatus was provided but no instructions beyond describing the theoretical expression and the apparatus. The point of the exercise is for the students to plan the measurement, deciding what to keep constant. This type of project would be suitable at a very early stage.

(b) The next level is to provide the apparatus but no theoretical expression and to ask the students to determine some empirical relations. I asked them to investigate the resolution of a bird-watching telescope as a function of distance from the object, aperture size and frequency of light. They were given a telescope, two lines on a screen, some filters and a measuring tape. This type of investigation requires much more thought and decision making than in (a). Students found it tough, partly because they had not been prepared properly for the task but my biggest issue was that the lab demonstrators struggled to cope with being facilitators as opposed to sources of information; instead of coaxing students to think, they just told them 'answers', which, to make matters worse, were often incorrect.

(c) Creating an artefact: I asked a pair of applied physics students to make a device that acted as an automatic, light-sensitive switch for me to use to deter burglars while I was away from home. An interesting point here was that it made a difference to the students that the project was based on a real need and not just an arbitrary exercise.

(d) On a similar principle, in the 1990s, when four-year physics degree programmes were introduced, I developed the idea of a group project, in which a team of students worked with an external sponsor from industry on a topic of genuine interest to the latter. Such arrangements are now more common and better organised but, even in those early days, it was evident how the external sponsorship made a positive difference to student attitudes. Anecdotally, the scheme was not popular with at least one of our external examiners, who was unhappy that the projects were not associated with leading-edge physics, a point that was robustly defended by me and my colleagues. However, as the scheme progressed, we did allow research groups to act as external sponsors but still using the basic principle of the projects, that the students took ownership of the work and were able to make decisions on how to go forward.

As an experimental physicist, I rarely set theory or modelling projects but I am aware that there are special problems around that type of investigation, not least that in the more advanced topics—string theory or cosmology spring to mind—it is hard to see how a first-degree student can do any research-level calculations. Sometimes, project sponsors deal with the issue by setting literature surveys but, while there is a place for literature summaries, they do not constitute projects in terms of the learning outcomes outlined above, and should not be used in that way. A cosmologist colleague produced a much more imaginative approach, setting a visualisation project, with a group of students working on a galactic simulator to be used both in research and outreach contexts.

I have recommended that students do not work alone on projects but as part of a pair or within a larger group. That raises the awkward question of whether to allow students to choose their own partners; there are undoubtedly arguments either way. My own view is that it depends on context: for a laboratory programme comprising a succession of mini-projects, I would put students into pairs/groups and change them for each task. An important skill is to be able to work with different types of people. For a longer, single project, I would ask individual students to choose from a list of projects and then pair or group them in the allocation process. I will return to this point later in the chapter.

5.2.2 Principles of assessing extended projects

To complete the section on projects, I concentrate on the more extended projects and consider what forms and principles of assessment would be appropriate to meet the criteria I have mooted as defining project work. I consider the mechanisms of the assessments in the next section.

- *Timed formative meetings.* The project sponsor should have regular, timed meetings with the students to answer questions and provide informal feedback. It is a good idea to have a brief, agreed outcome of these meetings written down to avoid any misunderstandings.
- *Project plan.* Depending on the scope of the project, this can include a literature survey, preliminary calculations or investigations and/or a design study. In my undergraduate final-year project, my lab partner and I showed that the project we had been assigned was literally impossible, which perhaps the sponsor might have realised but which also showed the value of the preliminary calculations.
- *Log/portfolio.* Students should be expected to keep a record of what they do, suitably dated. The easiest way to do this is to keep a physical, laboratory notebook, which has many virtues in that the format is never obsolete and the timeline is clear. But pens and paper are not in tune with our times so more likely is a computer-based log, where the times of the entries are noted. The log should form some part of the summative assessment; these days, professional physicists are required to provide their raw data on request and it is as well for students to get into good habits. One means of producing a record, which is highly suited to projects involving placements, is a portfolio of material, which can have both physical and virtual components.
- *Written report.* This can be in a number of forms. Two obvious ones are a journal paper, with formatting and length guidelines, or an internal report, such as one might find in industry. Whatever the format, students should be encouraged to be concise and to concentrate on the important points. Details have their place in the project log.
- *Oral presentation.* This element could be a short, conference-style presentation. A podcast is also possible, as are many other outputs, but it is essential that students are required to answer questions so, if the oral format does not allow that naturally, it may be necessary to arrange an interview.
- *Statement of contribution.* Some of the elements above can be joint (the project plan and possibly the log) while others, such as the written report would normally be individual. The project work itself is necessarily joint so each student needs to indicate what they and their colleagues contributed to the work. Evidently, such submissions should not have a direct input into the summative assessment but they provide very useful information for the assessors. They also provide an opportunity for students to indicate any issues and are therefore best made confidentially.
- *Variety of assessors.* The list above implies a variety of assessments but it is also a good idea to have a variety of assessors. I go further than that:

although the project sponsors should clearly have an input into the summative assessment, they should not have primary responsibility for it, nor should their mark constitute the major contribution. I am aware that this suggestion is counter to practice in most institutions but there are many advantages to it, most notably that it is much easier to produce consistency. Obviously, the sponsors need to have some input since they are the people who know best the students' contributions, but they have no means of comparing across the student cohort. In addition, project sponsors may be too close to the work to be objective—I have even known some sponsors be offended if a co-marker gives a low mark to *their* project. This effect may be exacerbated when the project is based in a research group.

5.3 Types of assessment

Returning now to more general aspects of investigative work, we revisit our desired outcomes, that is what we want the students to be able to do at the end of the programme. These need to be mapped onto assessment types, which, in turn, can be linked to types of activity. To some people, this may seem the wrong way around; they might feel that we should start with the activities and then find a way to assess them. The reason I choose to do it this way is because the real link is between outcome and assessment; student activity is driven by assessment. Once we know how the outcomes are to be assessed, we can choose activities to map onto the assessment types.

Table 5.1 is a grid linking the learning outcomes to various types of assessment. Later, I shall look at examples of the mechanisms and criteria for producing grades from these assessments. To take a concrete example, an assessment appropriate for the 'communicating results' outcome might be an oral presentation, and that is evident from the grid. However, that says nothing about the type of activity (project, set experiment etc) the student is undertaking, nor what methodology or criteria we use for awarding marks to the presentations.

Table 5.1. Matching types of assessment to learning outcomes.

Assessment \ Outcome	Illustrating physics	Health and safety	Familiarity with apparatus	Data handling	Investigation design	Scientific methodology	Scientific integrity	Keeping a record	Imagination, intuition, synopsis	Communicating results
Multiple choice questions		x	x	x	x					
Short written answers		x	x	x	x	x				
Written report, paper, blog etc.			x	x	x	x	x	x	x	x
Oral presentation						x			x	x
Oral examination		x			x	x			x	
Observation	x	x	x			x		x		
Numerical results				x	x					
Log/portfolio		x		x	x	x	x	x		
Design study		x	x		x				x	
Literature survey					x	x				
Podcast						x			x	x

Table 5.2. Matching activities to learning outcomes.

Activities \ Outcome	Illustrating physics	Health and Safety	Familiarity with apparatus	Data handling	Investigation design	Scientific methodology	Scientific integrity	Keeping a record	Imagination, intuition, synopsis	Communicating results
Examples/computing classes		x		x	(x)					
Set experiments	x		x	x				x		x
Mini-projects	x		x	x	x	x		x	x	x
Extended projects	x	x		x	x	x	x	x	x	x

As expected, no assessment is appropriate, or required, in order to illustrate physics, although it is a good idea to coordinate the physics covered in a laboratory course with the material covered in other modules. To clarify a couple of the entries in the table: 'observation' refers to an assessor watching students in action. It is the physics equivalent of a driving test. 'Numerical results' refers to a task in which students perform some type of measurement and are assessed solely on the quality of their results. The list of assessments is by no means exhaustive but it is clear that some learning outcomes are easier to assess than others. Scientific integrity is particularly elusive. Finally, I have included literature survey as a type of assessment, although it could be seen as an independent, stand-alone activity. The latter, however, which is essentially writing an essay, is not really a part of an investigative programme. To be so it must be linked to a subsequent investigation, which then makes it an element of assessment for that investigation rather than an activity in itself.

Table 5.2 links the types of assessment to a number of suggested activities. Examples classes, possibly involving computers, represent activities that are classroom-based, without requiring access to apparatus. Set experiments involve students performing experiments on pre-prepared apparatus according to a script. These are typically experiments that might last a few hours although there is nothing to exclude, for example, a carousel of relatively short measurements. Projects have been discussed in detail above but, to clarify, a mini-project is usually an investigation where, at the very least, students have some responsibility for the investigation; this description covers a range of possibilities from students testing a simple theoretical prediction using given apparatus, to them selecting the apparatus to investigate a more open-ended problem. In a theoretical context, the equivalent range might be from the adaptation of an existing model to do a calculation to the design of a model to solve a specific problem.

An important feature of a programme on investigative physics, and hence its assessment, is that it is coherent and that the various skills are developed with explicit teaching and are built upon as the course progresses. Typically, a programme would consist of stages, possibly, though not necessarily, linked to years, each stage involving activities linked to assessments and learning outcomes. There are many variations but an example is outlined below.

Stage 1

Activities	Possible assessments	Learning outcomes
Examples classes	MCQ Short answers	Data handling Health and safety
Set experiments	Log Brief report Observation Numerical results Oral examination	Familiarisation with apparatus Keeping a record Data handling Communication of results

Stage 2

Activities	Possible assessments	Learning outcomes
Mini-projects	Log Brief report Oral examination Oral presentation/podcast	Familiarisation with apparatus Keeping record Data handling Communication of results Investigation design Scientific methodology Imagination, intuition and synopsis

Stage 3

Activities	Possible assessments	Learning outcomes
Mini-projects	Log Brief report Oral examination Oral presentation/podcast	Familiarisation with apparatus Keeping record Data handling Communication of results Scientific methodology Investigation design Imagination, intuition and synopsis
Extended projects	Log/portfolio (including data store) Written report/paper Oral presentation Design study Literature survey (Statement of contribution)	Keeping record Data handling Communication of results Scientific methodology Investigation design Imagination, intuition and synopsis Health and safety Scientific integrity

In this model, stage 1 is introductory, explaining the essential features of investigative work to the students. In many ways, this can be the most challenging part of any programme, encouraging good habits without draining the exercise of any interest. One exercise I would discourage at this stage is the requirement to write a 'formal account'. The emphasis should be on learning the skills and not an arbitrary exercise of writing a report of dubious format. Instead, a brief, focused report should suffice. An alternative approach used successfully in several universities is to have a set of questions associated with each experiment and designed to make the students think about the experiment they have just completed.

I have included standard, set experiments in Stage 1 because that is the simplest way of introducing apparatus and ideas of measurement uncertainty. But these experiments do not need to last several hours. An alternative, often used in medical programmes, is to have a carousel of several, short experiments, with the emphasis on different techniques and analysis of measurement and uncertainty.

Stages 2 to 3 introduce project-based activity and assessment, initially with an emphasis on mini-projects, in which students put into place the ideas they have learned about the process of measurement, building to more open-ended investigations and culminating in an extended investigation as discussed in detail above. For stages 2 and 3, the reports become more sophisticated but should be based on a format that the student sees as sensible, for example a technical report or a scientific paper according to specific journal requirements.

There are, of course, many different models: the essential features are the linking to skills and the cumulative nature of the scheme. Later stages should build on the earlier ones. In some cases, the investigations might be tailored to a specific area of physics, such as electronics, interfacing, optics etc, but the principles remain the same. Alternatively, there might be an emphasis on theoretical investigations or modelling, where the familiarisation is not with apparatus but with standard techniques.

5.4 Examples of assessment criteria

Investigative work poses special problems for assessment. One advantage of a timed examination is that there is usually a single marker so that, once a model answer and a marking scheme have been devised, it is relatively easy to award marks consistently. Moreover, since the answer papers are usually anonymous, there is little possibility of personal bias. The assessment of investigative work is quite different. Frequently, a single task is marked by several assessors, of varying levels of experience, there is rarely a 'right answer' and certainly not for more advanced project work, and the person awarding the marks will know the identities of the students.

As a consequence of these problems, most universities have developed mark templates in which they attempt to provide a translation from performance to marks for the various assessment types described above. Examples of such templates are discussed in detail over the next few sections. Before embarking on that analysis, I wish to make a few general remarks, reinforcing some of the points I have made earlier. First, it is a feature of this type of assessment that the range of marks is compressed relative to that for a timed examination and that fail marks are relatively rare. I believe that this compression occurs for two reasons. The less positive one is that the markers do not have sufficient confidence in their objectivity to award fails or low marks. More positively, it really is the case that the gap between weaker and stronger students is less for continually assessed activities; my personal prejudice on that point is that students who do poorly in examinations often do so because they have not worked as regularly as they might have done, whereas, in investigative work, attendance is compulsory.

Another set of general remarks concern the marking of projects and double marking generally. By their nature, all projects are different and, in universities I am familiar with, most of their assessment is carried out by the project sponsor, although there is usually a second marker assigned. In my experience, this arrangement almost always leads to gross inconsistencies in marking, the consequences of which can be very serious since projects often form a significant fraction of the final year mark.

In my view, best practice is achieved by reducing the input of the project sponsor in the assessment process. Where the sponsor is external to the university, for example from industry, that usually occurs anyway, but not when the project is based in a university research team. The sponsor should have some input, certainly reporting attendance, inputs from the different students, the degree of help provided and so on. But the assessment of components such as a design study, a scientific paper or an oral presentation should be carried out by people who, while expert enough to understand the topic—projects are often very specialised—have not been directly associated with it. That is precisely the arrangement that many universities have for PhD examinations so it makes sense to use it for major projects.

Colleagues in essay-based subjects use double marking for all assessments, including timed examinations. In physics, the detail of the model answers makes this process not a good use of resources at what is a very busy time of year. However, it is essential that all substantial assessments for investigative work should be double marked. And double marking means that the two examiners assess entirely independently. That is, they are not aware of the marks awarded by the other person and that they submit their marks and reports to an independent, third person who decides if any moderation is required. Without such a procedure, it is unlikely that the marking will be truly independent, even more so if the project sponsor is one of the markers. I know from many examples that it is very difficult for a second marker to be independent if they can see the marks of the first marker. In addition, it is equally difficult for a junior member of staff to challenge the opinion of a more senior person.

5.4.1 Logs and performance

Modules associated with investigative work are primarily concerned with skills training rather than any specific knowledge base. In the early stages, those skills will comprise: familiarity with, and limitations of, standard measurement techniques; data processing, including the analysis of uncertainties; keeping a record; reporting results; and health and safety. The most important role of assessment in this context is to ensure that students acquire these skills.

Some of these skills, notable uncertainty analysis and health and safety considerations, could be tested using problem classes or independent multiple choice questions and I have used such approaches myself. However, my preferred approach is to try to link assessment to an experimental context wherever possible. Consequently, rather than having a problem class on, say, the combination of uncertainties, the technique is better assessed by designing a set of simple measurements that illustrate different aspects of the process. Similarly, rather than having a set of multiple choice questions on health and safety, it is preferable to give students different experimental set-ups and require them to think through the safety aspects. Of course, there will always be a need for general safety training before embarking on practical work.

I have mentioned earlier that there is absolutely no reason why marks should be added linearly in order to provide a final grade for a course. In an investigative programme there are elements where the important aspect is for students to show competence in a particular skill. For that, a *threshold* assessment is most apt. To progress, students must show competence in defined measurement techniques and in the handling of uncertainty, for example. Once the threshold is achieved, a uniform mark can be awarded and the students move on to the next stage. Such an approach is particularly suitable for health and safety.

The traditional first stage of a laboratory programme is to have a set of experiments, each of which is designed to last a single session. If done properly, the experiments can illustrate a range of physics (recall the relevant learning outcome) but also a range of data handling requirements, measurement techniques etc. This arrangement is fine and can work well; the danger is that students see it as handle-turning, all the more so if the experiments have very specific instructions that do not require students to engage their minds. That mindset is exacerbated if there is a need for a 'formal account' of some sort on some or all of the experiments. In 40 years of involvement in physics higher education, the most common moan from first year students is about having to do tedious formal accounts!

There are alternatives to these one-session, set experiments. One is to have a carousel of simple measurements—it is easy to think of examples—and students could be assessed in terms of the quality of their results and their corresponding uncertainties. Such an approach provides focus on what we want the students to be able to do and also has variety. Moving on from that, one can have mini-projects, which are essentially the same arrangements as the traditional set experiments but with students required to make decisions about the measurements instead of following a recipe.

For the carousel arrangement, the assessments can be threshold and comprise the actual numbers emerging from the measurements. Once these have been approved, possibly during an interview with a demonstrator, students can progress to the next stage. For the mini-projects, students can be assessed in terms of a number of diagnostic questions that guide decisions (e.g. 'Is it better to vary current or voltage?'; 'What constitutes the principal source of uncertainty?'; 'What assumptions are you making'; and so on), and on the quality of their results and analysis.

In practice students are rarely assessed directly on their ability to carry out investigations. Instead, it is common to use proxies, such the results obtained in set experiments, the log, or via a formal account, the latter being the least informative.

Table 5.3 shows an example of a marking sheet for a laboratory demonstrator to use in a first year laboratory that uses set experiments. Demonstrators may be academics or postgraduate students. The marks are to be awarded on the basis of the assessor's observation of the way the student carried out the experiment and an interview that takes place at the end of the session, in which the students are asked a series of questions to test their understanding of what they have done. There is also a grade for the quality of the results. The criteria for the assessment are listed in table 5.4, which is discussed below.

Table 5.3 represents a laudable attempt to assess actual performance instead of using a proxy but there are disadvantages. The most important one is that the interview time is time taken from the session and, given that one demonstrator covers several experiments, students will generally have to wait around until it is their turn. In addition, one imagines that the number of questions that one can ask about a particular set experiment, where students follow a recipe, is fairly limited so that students can pass them on to their friends and the marking loses its validity. On a more pedantic level, the sheet would be improved by using the word 'uncertainties' rather than 'errors', however common the latter term is. There can be errors in carrying out an investigation, but they are not the same as the uncertainty with which we know the results.

Table 5.4 shows the criteria to be used by the demonstrators to inform their grading. It is remarkably fine-grained in the first-class category; I am not sure of the reasoning for this level of detail but it might well bias demonstrators on the balance of the grades they give. We are going to meet several examples of this type of grid in this and later sections and there are three general remarks, two closely linked, that apply to all of them to a greater or lesser degree.

The first remark concerns the definitions themselves and table 5.4 is actually better than most in this respect: it is that the criteria should describe achievement, not the lack thereof. In the current example, this is the case down to grade C, lower second class but the grade descriptor for a third class performance comprises almost entirely negative words: 'disorganised', 'unsystematic', 'only partial', 'difficulty', 'passively', 'inadequate', 'confused', 'limited', 'unable'. What a grim picture that paints of an Honours degree student.

The second point is the arbitrariness of the words used to define the various grades. What exactly distinguishes 'exceptionally', 'excellently', 'very well', 'well' and 'adequately'? One might contend that the last is merely turning the handle but the rest are arguably synonyms. Elsewhere, we have students potentially displaying 'outstanding', 'thorough', 'clear' and 'good' understanding. Without further definition, these words offer little genuine guidance on absolute standards.

The third point follows directly from the second and refers to a theme of this book. These definitions make sense only if one accepts that one is norm referencing. Indeed, words such as 'exceptional' and 'outstanding' imply a comparison with

Table 5.3. Example of assessment of performance.

First Year Laboratory Physics

Student Performance Evaluation Form

Student Name: _____

Experiment title: _____

Date: _____

Briefly evaluate student's performance for this experiment

Preparation:

Interest in work/Lab performance:

Results:

Analysis/Errors:

Answers to questions:

Overall (from 10):

Please briefly describe student's performance with few sentences:

Demonstrator:

Table 5.4. Criteria for assessing performance in laboratory work and oral interview.

Grade	Descriptor
First class **A++** **>90%**	• Exceptionally well prepared, displaying a systematic and carefully planned approach with a comprehensive understanding of the material and methodology. • Able to work independently, or to take a leading role in a group. Outstanding presentational skills showing an accurate and fluent analysis of the topic or problem. • Answers questions thoughtfully and accurately with independent ideas. Able to reach valid/relevant/perceptive conclusions and to suggest logical and original extensions of the work.
First class **A+** **80%–89%**	• Excellently prepared, displaying a systematic and carefully planned approach with a thorough understanding of the material and methodology. • Able to work independently or to participate effectively in a group. • Excellent presentational skills showing an accurate and fluent analysis of the topic or problem. • Answers questions thoughtfully and accurately with independent ideas. Able to reach • valid/relevant/perceptive conclusions and to suggest logical and appropriate extensions of the work.
First class **A** **70%–79%**	• Very well prepared, displaying a systematic and carefully planned approach with a clear understanding of the material and methodology. • Able to work independently or to participate constructively in a group. • Very good presentational skills showing an accurate and fluent analysis of the topic or problem. • Answers questions thoughtfully and accurately with independent ideas. • Able to reach valid/relevant conclusions and to suggest logical extensions of the work.
Upper second class **B** **60%–69%**	• Well prepared, displaying a systematic approach and a good understanding of the material and methodology. • Able to work independently or to participate actively in a group. • Good presentational skills showing a fluent analysis of the topic or problem. • Answers questions thoughtfully with some independent ideas. • Able to reach valid/relevant conclusions and to suggest some logical extensions of the work.
Lower second class **C** **50%–59%**	• Adequately prepared, displaying a reasonably systematic approach and some understanding of the material and methodology. • Able to work independently or to participate in a group. • Adequate presentational skills showing a credible analysis of the topic or problem. • Answers questions with some wider understanding of the key ideas.

(*Continued*)

Table 5.4. (*Continued*)

Grade	Descriptor
	• Able to reach valid conclusions and to suggest extensions of the work.
Third class **D** **40%–49%**	• Disorganised preparation, displaying an unsystematic approach, and only partial understanding of the material and methodology. • Has difficulty in working independently or participates only passively in a group. • Inadequate presentational skills showing a confused analysis of the topic or problem. • Answers to questions show limited understanding of the key ideas. • Able to reach some valid conclusions but unable to suggest appropriate extensions of the work.
Fail **F** **30%–39%**	• Poor preparation, displaying an unsystematic approach and very limited understanding of the material and methodology. • Has difficulty in working independently or participates ineffectively in a group. • Poor presentational skills showing a confused analysis of the topic or problem. • Answers to questions show little understanding of the key ideas. • Unable to reach valid conclusions or to suggest appropriate extensions of the work.
Fail **F** **10%–29%**	• Very limited preparation with no understanding of the material and methodology. • Has great difficulty in working independently or unable to work in a group. • Almost no presentational skills with no analysis of the topic or problem. Answers to questions show almost no understanding of the subject. • Unable to reach any relevant conclusions.

other students in the same cohort. That is not to say that the particular criteria listed in table 5.4 are deficient; they are typical across many universities. Their worth is that they make clear to students what they are being assessed on. However, given the arbitrariness of the definitions, they are likely to lead to substantial inconsistency if applied by independent markers. At more advanced stages, students will be carrying out project work, which means that there will be much less consistency between student activities and, in many cases, students work independently so there will be no demonstrators to monitor performance. Traditionally, students undertaking projects were expected to keep paper laboratory logs but few universities nowadays require this approach. Nevertheless, there is a need to assess how well the project has been carried out, how decisions were made and so on. Some of these points have been discussed previously but one way of addressing the issue is to use a *portfolio* approach.

The example shown in table 5.5 is interesting for a number of reasons. It refers to the scientific investigation element of a wider set of activities, which include problem sheets and group work, and explains to students what is required to pass; in other words, it defines a minimum *threshold* performance for satisfactory progression. In addition, there is a list of the relevant learning outcomes. The table only makes sense in the particular context of this university but its approach is more generally applicable. Note that this table does not attempt to define what is meant, for example, by 'satisfactory', which requires a grid similar to that of table 5.4, with its attendant virtues and deficiencies.

Table 5.6 is equally novel and refers to the provision of formative feedback on submissions of draft portfolios. Rubric descriptors 1 and 2 refer to different stages in the investigation. I return to the portfolio approach in chapter 6 in the context of a more holistic approach to assessment at the programme level.

In the later stages of programmes, most investigative work is in the form of projects, which, as discussed above, pose special problems for assessment due to the diversity of the student experience. An attempt to provide guidance for the assessment of the performance side of a project—as distinct from the presentational aspects—is provided in table 5.7. I have lightly edited the original for reasons of space but have not made any significant changes.

There is much to applaud in this attempt to standardise assessment: it highlights the important points for students and tries to be as objective as possible in the grade descriptors. In the project discussion above, I made the point that project supervisors should be involved in the assessment process as little as possible but it is inevitable that, in considering the performance aspects, it is the supervisor that has to make the judgements. Consequently, this is a grid for project supervisors to use. Note also that there is reference to a 'notebook' which is presumably paper based.

On the negative side, the grid does fall heavily into the trap of starting at the top in defining excellence and then describing the other grades in various degrees of the student's inability to perform at the highest level. Even the 60% to 69% category, corresponding to a very good degree, has negative words such as 'reasonable', 'not fully comprehensive', 'some evidence', 'requires supervision', 'needed advice' and so on. The descriptors for 50% to 59% are almost entirely negative. Since the grid is designed for a module with a pass mark of 50%, it is reasonable that the lowest category refers to failure, although the descriptors are so negative that it is hard to see why such a performance should be awarded any marks.

The only grades that are described in uniformly positive terms are the top two but here we see a repeat of the general issue identified earlier, which is how to distinguish between descriptor elements that are essentially synonymous. For example, 'Results are performed [*sic*] at the optimum level' is contrasted with 'Results are consistent with those expected from a skilled first year PhD': in my dictionary, the word optimum means the best possible, so the descriptor for the lower grade, by definition, cannot be improved upon. In fact, the descriptor for the higher grade relies on an understanding of what we mean by a 'skilled PhD student'.

Table 5.5. Portfolio approach to assessment of scientific investigation.

Learning outcome(s) assessed	Components	Required for pass	Final rubric descriptor for 'fail'
Plan and execute closed and open scientific investigations with a consideration for ethics in science. Including:	Formative poster (A2) + feedback Formative lab report (8 pages)	Submission of Fourier poster to satisfactory level AND/OR lab report on partially scripted investigation to satisfactory level.	One or more of the following not present: Fourier poster satisfactory OR lab report satisfactory
Select, critically evaluate and use appropriate resources for an investigation.	Project plans (1–2 pages)	Project plans included for at least one of partially scripted investigation and/or research project—must be signed by lab staff.	1 × signed project plans
Manipulate, systematically analyse and interpret scientific data.	Risk assessments	Risk assessments included and signed by staff member for experimental work.	2 × signed risk assessments
Present information accurately and efficiently with due consideration of the audience.	Investigation summary (200–400 words)	Summary of steps that the student has taken during their investigation(s)—can relate to any project(s)—must show how they have used appropriate resources and applied their knowledge. Can include scans of lab books, feedback on assignments, bibliography etc.	Investigation summary
Synthesise and apply physics knowledge and investigative skills to physics problem solving and authentic challenges.	Discussion of ethical considerations (50–150 words)	Narrative of ethical considerations—can relate to any project(s). If not specific to a project, include a narrative of ethical considerations in general and how these apply to scientific research.	Ethical considerations narrative

Table 5.6. Formative marking scheme for draft portfolio.

Check point 2 rubric descriptor for 'needs attention'	Check point 2 rubric descriptor for 'on target to satisfactory'	Check point 2 rubric descriptor for 'on target to excellent'	Check point 1 rubric descriptor for 'needs attention'	Check point 1 rubric descriptor for 'on target to satisfactory'	Check point 1 rubric descriptor for 'on target to excellent'
One or more of the following is not present: Fourier poster satisfactory mark 1 × project plans AND risk assessments Attempt at an investigation summary Beginning of an ethical considerations narrative	Student has shown that their Fourier poster has been submitted to a satisfactory standard (as evidenced in attached feedback given by lab teachers). At least 1 × project plans and risk assessments are included and signed by lab staff. The investigation summary shows that the student has selected and used appropriate resources but may not be fully complete yet. There is at least some narrative of ethical considerations relating to at least one project or to wider ethical considerations in physics in general.	Satisfies criteria for on target to satisfactory. In addition both of the following apply: The Fourier poster is awarded excellent. Investigation summary is detailed and includes the use of well selected and high quality resources that have clearly steered the investigation AND/OR the narrative of ethical considerations is very well thought out and relates to the student's own work.	One or more of the following is not present: Beginning of an investigation summary. Beginning of a narrative on ethical considerations.	The student has begun to write an investigation summary for one of their lab experiments AND/OR has begun their narrative on ethical considerations.	Satisfies criteria for satisfactory. In addition the investigation summary is well underway for one of their lab experiments AND for the narrative on ethical considerations.

Table 5.7. Assessment grid for performance aspects of project.

Aspect	<50%	50%–59%	60%–69%	70%–84%	85%–100%
Quality of the research carried out	No results or results meaningless due to failure to apply the scientific method; student damaged equipment or worked in an unsafe manner. Notebook contains little or no information relating to experimental work done.	Some results but limited due to poor use of equipment/ technique /method. Notebook includes only some of the critical points to reproduce work. Results included as undated loose pages, and/or data not recorded in a safe environment.	Results are reasonable for the given facilities (equipment/ code etc) but not necessarily optimising what was available. Notebook contains most parameters and evidence of key analysis with dates but is not fully comprehensive.	Results are performed at the optimum level (consistent with facilities provided). Notebook contains full details of experimental parameters, dates, data, methodology and results analysed.	Results are consistent with those expected from a skilled first year PhD with similar facilities. Notebook provides rigorous trail of parameters, methodology and data. It also contains critical views of data with observations and theories to investigate.
Critical faculties and independence	Student did not demonstrate any degree of critical thinking even when prompted, did not take action on own initiative or when told to do so. No engagement in critical discussion with the supervisor.	Student demonstrated limited critical thinking when prompted. Student did not work independently of demonstrator. Minimal engagement in critical discussion with the supervisor. Prepared to solve	Student demonstrated some evidence of ability to think critically. Main results are analysed with appropriate theory/models with uncertainties. Some engagement in critical discussion with the supervisor.	Student applied independent critical judgement considering results. Help needed only while learning new techniques. Results analysed within the context of literature and make use of uncertainties as	Added knowledge by independent work/ analysis applied to the project. Critical judgement shown in the interpretation of results beyond discussions with supervisor. Creative engagement in discussions with the

Criterion				
	No effort made to solve problems even with assistance.	problems only with direct supervision, unable to diagnose problems independently.	Independently diagnose problems, but requires supervision to solve problems.	required. Proactive engagement in discussion with the supervisor. Diagnosed and corrected problems as they arose.	supervisor. Problems diagnosed and solved independently, with improvements to technique/methodology investigated.
Overall project planning and management	Student has failed to complete activities, failed to turn up for meetings, was absent without good explanation. Disruptive use of infrastructure.	Student has wasted time and/or failed to complete key activities without good reason. Student was late for meetings without good explanation.	Student has managed to complete most tasks. Student has needed advice to set reasonable timelines.	Student completed the required tasks for the lab, managing time well. Project conclusion has been obtained, coherent with the task(s) engaged.	Student has set realistic deadlines and timescales, prioritised activities and reached a project conclusion beyond expectations. Optimal use of infrastructure.
Extension of project beyond initial set goals	Student failed to complete most of the set tasks, let alone extend the work.	Student completed only the more straightforward tasks without extending the work.	Students completed set tasks but did not extend project significantly.	Students managed some degree of extension beyond set tasks.	Student independently devised extension to project.

As a final point on performance, group projects, where students work together on a problem, are highly desirable in any degree programme because students are encouraged to talk to each other about physics. Generally, students will be given information and training on how to work as a group. The corresponding assessment necessarily includes a consideration of the performance of the group as a whole; in some cases, that might comprise the totality of the marks, although it is more common also to include some sort of individual assessment. Either way, students are often unhappy about the perceived fairness. In some cases, they may feel a fellow student has been freeloading, or it may be that another student is withholding information to the rest to try to gain an advantage.

To overcome such concerns, it is best to allow students to comment on the relative contributions of their peers. An extreme approach, which I have tried unsuccessfully, is to award a project mark, say 60% and multiply that by the number of people in the group, say 5, to give 300 marks. Then the students allocate the marks themselves. The reason this method failed in my case was that the student groups all ended up offering equal marks, even though privately individual students expressed their concerns.

A more considered methodology first asks each student to produce a brief narrative in which they state their own contribution and also those of their colleagues. This is a useful tool for the assessor to see the consistency of the responses. Second, each student may be asked in confidence to produce a more quantitative assessment of each person's contribution. An example of such an input is shown in table 5.8. I am unsure whether, in this case, the submitting students should include their own name, but I think it is good practice to do so.

Table 5.8 is straightforward enough. I note that the variations allowed offer more space below the average than above, perhaps indicating that the main purpose of the form is to unearth those not contributing sufficiently. In addition, the shading and the requirement to justify any departures from par give a strong push to students to indicate equality of marks.

Table 5.8. Student input on relative performance.

	0%	-20%	-10%	Equal	+5%	+10%
Name 1						
Name 2						
Name 3						
Name 4						
Name 5						
Name 6						
Comments to back-up unequal weightings:						

Even allowing for this type of peer contribution to assessment, it is clear that peer marks cannot be directly included in the overall assessment. However, they can be used as indicators of a problem in the team. Ultimately, it should be the team's performance as a whole that is assessed, but students must have some means of indicating that their work might have been affected adversely by issues with a colleague.

In summary, it is not easy to assess performance in investigative work. At the early stages, where students are likely to be carrying out set exercises or, at least, tasks that other students have done in previous years, it is relatively easy to determine whether they have performed well relative to their contemporaries. However, once the investigations have a degree of independence and decision making, comparison is much more difficult. My suggestions for good practice are:

- Concentrate on the learning outcomes, principally (but not exclusively):
    - Keeping a record.
    - Data handling.
    - Communication of results.
    - Scientific methodology.
    - Health and safety.
- Provide clear descriptions of what is required to students.
- Use relatively few grades: for most elements, it is sufficient to indicate whether a student is below par, par or above par for a particular task.
- Do not be obsessed by grading. For many tasks, for example the successful use of a piece of apparatus, a threshold assessment is sufficient.

This last bullet is worth considering in a little more detail. The assessment of performance is never going to lend itself readily to a full percentage scale because much of the teaching of investigative skills is training in competence, particularly in the early stages, with relatively narrow distributions of performance. The enemy here is grading by the simple addition of marks from various elements of assessment to provide an overall mark for the course. It is perfectly possible to provide an overall grade or mark by other means. Such methods will be discussed in a more general context in chapter 6.

5.4.2 Written and oral presentations

Any piece of investigative work has to be reported in some manner and, due to the issues around assessing the actual student activity, universities usually have most marks on offer for the reporting aspects. With apologies to the reader for more of the same, in tables 5.9 and 5.10 I reproduce two marking grids of my own creation for oral and written presentations, respectively. Although these were developed 30 years ago for modules dedicated to improving communication skills, they apply to generic presentations, including those specific to investigative projects. Indeed, checking how project presentations are assessed across a range of universities, I note that similar grids are in use.

Table 5.9. Assessment guidelines for oral presentations.

Item	Outstanding	Good	Satisfactory	Poor
Introduction	Clear introduction with central theme and purpose stated.	Introduction gives the basic theme.	Introduction does not match talk or is too brief/long.	No introduction.
Science	Good, clear, imaginative explanations and error free.	No errors of science and consistent explanations.	No errors and adequate explanations.	Errors of science and/or explanations unconvincing.
Logical flow	Each section is logically coherent and links to other sections and the central theme. Good flow with evidence of planning.	Sections are logically self-consistent.	Generally argument is logical with perhaps a few discontinuities.	Haphazard and disorganised.
Quantity of content	Quantity is well matched to the time available. The talk does not sound rushed and does not have unnecessary detail.	Quantity fits the time available and is well-structured.	The talk is completed on time.	The quantity of material is inappropriate for the time available.
Matching to task	The talk matches perfectly the task in terms of content, level and identifying the target audience.	The target audience is judged correctly and the level well judged.	Talk matches task and is reasonably linked to the target audience.	The talk misses the target.
Summary	It is clear that the talk has ended. There is a cogent summary of the key points. Where relevant, there is an indication of future work.	A clear summary is provided of the main conclusions which link to the points made.	There is a summary.	There is no summary.
Audience rapport	Plenty of eye contact with audience and positive body language. The speaker always looks at the audience when speaking to them. The audience is	Eye contact is maintained and the speaker has good posture, standing in a good position and largely	Eye contact generally maintained and no long periods looking away. Few tics.	No eye contact and poor body language. Words or phrases repeated and/or physical mannerisms.

	engaged. The speaker has no verbal or physical mannerisms.	avoiding verbal or physical mannerisms.	
Enthusiasm	The speaker is enthusiastic about the talk and the subject; they sound as if they care about the audience understanding.	Speaker seems to be interested in the material and the audience.	Speaker maintains an interest in the material and links to the audience.	The speaker seems to wish s/he were somewhere else.
Visual aids	Visual aids are clear, informative and well-chosen. Slides are not cluttered, make good use of colour and are clearly visible to the whole audience. They add to the talk.	Visual aids are clear and well-chosen. They are visible to all and relevant to the material.	Visual aids are clear and visible to all.	Visual aids are missing, unhelpful, irrelevant or cannot be seen clearly.
Voice modulation	Both the tone and the volume of the voice are used for emphasis. There is a variation of pace to avoid monotony.	The talk is delivered clearly with some modulation of voice, tone and pace.	The talk is delivered clearly with some modulation of volume, tone or pace.	Largely a monotonous tone with no variation.
Question handling	Positive response to questions. Questions are answered with confidence and authority. The answers match the question and are at the right level.	Questions are welcomed and answered accurately with reference to the talk.	Questions are answered correctly.	Questions are not answered and/or not taken seriously.

Table 5.10. Assessment guidelines for written presentations.

Item	Outstanding	Good	Satisfactory	Poor
Introduction	Clear introduction with central theme and purpose stated.	Introduction gives the basic theme.	Introduction does not match topic or is too brief/long.	No proper introduction.
Science	Good, clear, imaginative explanations and error free.	No errors of science and consistent explanations.	No errors and adequate explanations.	Science errors and/or explanations unconvincing.
Logical flow	Each section is logically coherent and links to other sections and the central theme. Good flow with evidence of planning.	Sections are logically self-consistent.	Generally, argument is logical with perhaps a few discontinuities.	Haphazard and disorganised.
Quantity of material	Quantity is well matched to the number of words allowed. The article increases the knowledge of the reader without unnecessary detail.	Quantity fits word limit, has no obvious omissions and is well-structured.	The article fits within the word limit.	The quantity of material is inappropriate and/or there are major omissions.
Matching to task	The article matches perfectly the task in terms of content, level and identifying the target audience.	The target audience is judged correctly and the level well judged.	The article matches the task and is reasonably linked to the target audience.	The article misses the target.
Clarity of language	Language is clear, economic and unambiguous. Sentence structure is simple, vocabulary matched to audience and there is no jargon. Uses active voice where possible and has a consistent style. No undefined abbreviations.	The language is clear and unambiguous, with sensible sentence structure. Vocabulary is matched to the audience.	Mainly clear and unambiguous. The reader can obtain good comprehension of what is being said.	Several places where the language is ambiguous and meaning is difficult to comprehend.

Grammar	English is essentially error free, with proper use of definite and indefinite articles, verbs agreeing with subjects, adjectives agreeing with nouns and adverbs agreeing with verbs. All sentences have active verbs.	English is essentially error free to the extent that the meaning is preserved. No major grammatical errors.	Few grammatical errors and none that affect the meaning in a serious manner.	Several errors that make the article difficult to understand.
Punctuation	Punctuation is clear, simple and consistent. Correct use of commas, colons, semi-colons, apostrophes and full-stops. Sentences and paragraphs are correctly delineated.	Punctuation is clear, simple and consistent and leads to no ambiguities of meaning.	Any errors in punctuation do not lead to serious ambiguities.	Punctuation is sufficiently poor to lead to serious ambiguity and/or loss of readability.
Diagrams	Diagrams are well-chosen, necessary and informative. They are clearly presented, labelled Fig 1 etc, and have brief, informative captions. They are referred to from the text.	Diagrams are clear and useful and have labels and captions.	Diagrams are clearly presented and have a purpose in the text	Diagrams are missing when necessary. Diagrams are not labelled or presented poorly.
Referencing	There are references to all material which is not the original work of the author. References are presented in a standard format and may be unambiguously determined from the information given.	There are references to all material taken from other sources.	References are included and there is nothing used from another source that is not referenced.	There is inadequate referencing. **Note that insufficient referencing may lead to a charge of plagiarism.**

I should say that these grids were not the only information provided to students about how to give talks or write reports, articles etc. There was a comprehensive document covering all aspects of the presentations, including content, style, vocal intonation etc. There were also a number of tutorial sessions, plus the involvement of a professional communicator. I mention these points because, in some cases, students are somehow expected to work these matters out for themselves. While it is certainly true that assessment drives activity, in dealing with presentations, where small technical issues can make a large difference, students should not be expected to work it out for themselves.

I do not present these grids as particularly superior to those we have seen already. One good point, I think, is that the number of categories is relatively small and are not linked directly to numerical marks. Evidently, there does have to be a method of converting grades into marks and students should be aware of it. I shall return to this point in chapter 6 but for the moment, I make the point that one can argue that a linear addition of marks for the various elements of, say, a talk is undesirable. For example, if a student mumbles all the way through, with her or his back to the audience, it does not matter how erudite the content is or how attractive the slides: the talk is a fail.

Returning to the grids, I realised that, by and large, I had managed to use positive statements to define the various pass grades. There are exceptions—the 'Satisfactory' statement on the 'Introduction' is wholly negative—but generally, even the 'Satisfactory' criteria can be read independently as worthy achievements. The 'Poor' category corresponds to fail and is written in negative terms. To avoid being too kind to myself, I note that the 'Outstanding' category criteria tend to be longer than the others, perhaps indicating the usual bias to seeing anything less than the top grade as somehow disappointing.

Another trap I tried to avoid, with limited success, was the use of words such as 'outstanding', 'excellent', 'very good' etc, in descending order of quality, which we saw in some of the other grids.

Instead, I tried to be more specific about what constituted 'Outstanding' etc. I have not entirely succeeded, particularly in the grid for written presentations, because there are still words such as 'adequate' to interpret and also some grade distinctions that, at best, require judgements. What I have not been able to do is to remove entirely the norm-referenced character of such grids. We mark what we have in front of us; that does not mean that these grids serve no purpose but it does mean that they provide no absolute benchmark.

I will mention a couple of issues I have encountered, which are actually related to the absence of an absolute benchmark. The first time I developed the grids, in the 1990s, all the students being assessed were products of the UK school education system and had English as their first language. When I revisited the grids in 2017, the class mainly comprised students who had English as a second or third language. Inevitably, and entirely understandably, the quality of their written English was not as good as that of the two or three people who did have English as their mother tongue.

The other issue was the presence of dyslexic students in the class. In a timed examination, the students are given extra time, usually on the recommendation of a disability advisor, and that works well. For an essay, it might be possible to negotiate a delayed submission date. Neither possibility was viable in this case: there was no examination and delays to submissions to written work would have eaten into the preparation time for the other exercises. The solution I used was to work closely with the students involved to help them with their timetabling and in using advisors to develop coping strategies. Generally, the students involved were very positive and saw the course as a necessary challenge. However, I remained uneasy and uncertain about the fairness of the procedure.

Extensive and timely feedback is essential for the smooth running of any skills training if students are to improve. For the written presentations, with colleagues, I produced annotated notes plus an overall feedback narrative. The presentations were relatively brief at 500 words but I am aware that this is a significant marking burden, albeit compensated by the lack of the need to mark examination papers. For the oral presentations, I used a feedback form similar to that shown in table 5.11, which was designed to be used in conjunction with table 5.9. The same forms were also used for self-assessment—talks were recorded, a feature which also preserved a record of the event—and for peer assessment, with other students in the group completing forms for their colleagues. I did not use the peer feedback for any summative purpose; its main rationale was to ensure the watching students were concentrating on the important elements to help them with their own preparations.

Written and oral presentations are universally employed as assessment tools for practical work, in particular projects, often forming the lion's share of the marks available. In some sense, they act as proxies for the actual project work: it is possible to give an excellent talk on a poor project and vice versa. It follows, therefore, that students who are being assessed in this way must be trained in making these sorts of presentations. Otherwise, the project mark will be skewed away from the performance of the project towards the competence of the presentations. One has to consider carefully what skills are being assessed.

5.4.3 Other types of assessment

The most common additional or alternative assessment for investigative work is the poster. It offers a different experience for students and can be incorporated into a class poster session, with students answering questions about their work and where peers and academics can vote for the posters that they see as the most appealing, most informative etc. In terms of assessment, posters are compared against (yet another!) grid. Table 5.12 shows a (slightly edited) example for a module concerned with Fourier analysis.

The grid includes the desired learning outcomes and, commendably, allows only three classifications of achievement, an approach that assists in avoiding the trap of trying to distinguish the subtle differences between 'outstanding', 'excellent', 'very good' and so on. However, I think the designers started with the 'Excellent' column and then thought about departures from that level: the 'Satisfactory' column is

Table 5.11. Feedback form for oral assessment.

ORAL PRESENTATION FEEDBACK

Presenter: **Title:** **Date:**

Item	Grade (O,G,S,P)	Comments
Introduction		
Science		
Logical flow		
Quantity of content		
Matching to task		
Summary		
Audience rapport		
Enthusiasm		
Visual aids		
Voice modulation		
Question handling		
General comments		

Table 5.12. Assessment grid for poster presentation.

Learning outcome	Criteria	Fail	Satisfactory	Excellent
	Overall achievement	Student did not meet the satisfactory criteria in at least one of the categories—see comments.	Student has met the satisfactory criteria for all categories and can add this to their portfolio.	Student has exceeded the satisfactory criteria and achieved excellent results
Present information accurately and efficiently with due consideration of the audience.	Presentation	The structure makes the poster incomprehensible or the poster is missing at least three of the relevant sections: abstract, introduction, methods, diagrams, results & analysis, discussion, conclusions, references, diagrams.	The poster contains most of the necessary elements and includes diagram(s)/figure(s) to aid the understanding of the work. The use of space and logical structure is satisfactory but may have either too much space, too many words or poor use of diagrams.	The layout makes excellent use of the available space with good use of properly labelled and captioned diagrams. The content is not too wordy and flows well with a logical structure, title, summary, all relevant sections, and correctly presented references.
Apply Fourier analysis to optics and other physical systems.	Content 1: theory	Either there is no theory presented or the theory is incorrect, does not make sense or is not relevant.	Theory has been presented that provides some background context for the discussion. There may be some minor errors or inconsistencies but is on the whole correct.	The theory presented is comprehensive, correct and provides excellent background for the work and the discussion of the results refers to the theory presented. Theory has been taken from lectures and from referenced external sources.

(*Continued*)

Table 5.12. (*Continued*)

Learning outcome	Criteria	Fail	Satisfactory	Excellent
Manipulate, systematically analyse and interpret scientific data	Content 2: methods	Either no methods are presented or there are significant errors in the methods or they do not clearly describe the process involved.	Experimental methods are mostly correct and describe the relevant processes with a useful diagram if appropriate.	Experimental methods are clear, concise and well written, describing the relevant processes with a useful, well labelled diagram that aids the understanding. There are no omissions in experimental procedure or equipment.
	Content 3: results	Results are either missing or there are significant omissions such as units and errors or they are not presented in a suitable format.	Results are presented in a suitable format but may be lacking in clarity or missing occasional units, errors or labelling.	Experimental results are well presented in a suitable format with units present and errors quoted to the correct precision. Any figures are correctly labelled.
	Content 4: analysis and discussion	Analysis is either missing or an attempt has been made to analyse the data but it is not correct or cannot be followed. There may be little to no discussion of the results.	Data have been analysed with an attempt to justify uncertainties and errors. Analysis method can be followed relatively easily.	Data analysis is correct, clear and easy to follow and includes proper treatment of experimental uncertainties and relates to theory presented.

perhaps presented more negatively than is ideal, referring to absences, minor errors, lack of clarity etc. In addition, with a nod to my usual comment about norm referencing, in some of the entries there is a circularity; for example, one of the descriptors for excellence is 'The theory presented ... provides excellent background for the work'. A minor point is that the 'Criteria' column is probably mislabelled, since the elements of the 'Fail', 'Satisfactory' and 'Excellent' columns constitute the criteria. As in the other grids, the main virtue is to highlight to students what is important about the exercise.

Another common form of output for investigative work is the video or podcast. In essence, there are refinements of an oral presentation, incorporating technical aspects but without the pressure of having to deliver a live talk. In some cases, perhaps in helping a student with a disability or even when a pandemic is preventing in-person interactions, such devices can be used as alternatives to oral presentations. There is an argument that this type of output is becoming increasingly common in science, as well as in the world at large, and students are not only comfortable producing such artefacts, they also enjoy doing so.

I will spare the reader yet another grid: table 5.9 can be modified simply enough, including the addition of various technical aspects, sound as sound quality, continuity, lighting and so on.

5.5 General remarks

Investigative work is an essential part of any physics programme and it is the area where, currently, most of the skills development takes place. In designing a set of modules or courses to enable this development, there are two points to bear in mind. First, there must be a coherent progression of skills, culminating in some sort of project. Second, assessment should be an integral part of the course design.

To expand the second point, the starting point must be the investigative skills and knowledge that the students are expected to have on completion, the learning outcomes. Having identified those learning outcomes, the next step involves deciding how you will know whether the students have achieved them and that process requires the identification of the assessments to be used, alongside the various tasks and activities to be assessed.

It is inevitable that the early stages of investigative work involve training in the basics of scientific methodology, including familiarisation with standard apparatus, keeping a record, data handling, uncertainty analysis, health and safety and so on. Since the expectation is that all students should reach competence in these outcomes, it makes little sense to award marks on a 0%–100% scale, with a linear addition to produce an overall mark. A much more logical approach is to have a competence-based progression, in which students achieve a threshold of competence and then move on the next stage. That arrangement gives students a sense of progression—a common complaint about laboratory work is that it is monotonous—and provides a higher degree of assurance that the skills are embedded. In the system of awarding marks, a student with a low mark has probably not displayed the relevant learning outcome but is still allowed to progress, possibly to become demoralised at a later stage.

There are various ways in which threshold competence may be displayed. For the use of apparatus, it might be as simple as being asked to make a set of measurements and the values obtained are a measure of how well the equipment has been deployed. Health and safety, an area where competence must be guaranteed, can be assessed using quizzes, prepared experimental arrangements and/or focused question interviews.

An obvious drawback to this approach, relative to the awarding of marks on a 0%–100% scale, is that it is difficult to award a mark for the particular module if everyone eventually achieves competence in the learning outcomes. There are several ways of dealing with this issue; a few are suggested below:

- An extreme line would be to make the module a training course which must be passed but either does not contribute to the overall mark for the year, or provides the same mark for each student. Although this makes logical sense, it is unlikely to be found acceptable by university administrative centres and may also lead to some students doing the minimum work needed.
- At each threshold, students could be rated as fail, satisfactory, outstanding. At the end of the process, all progressing students would have a collection of satisfactory and outstanding grades, which could be converted to an overall mark.
- There could be a final section of advanced thresholds, perhaps with less explicit guidance. None of these tasks could be absolutely essential for progression but would allow students to progress to different degrees and marks could be awarded for how far a student has progressed.
- A refinement of the last bullet could be a mini-project, in which students display to what degree they can employ the competences learned.

A feature of investigative assessment that often concerns people is that the marks are more clustered than would be the case for a typical timed examination in physics. The suggestions above would hardly change that situation. However, I do not see it as a problem. Often, the reason people do poorly in examinations is because they have struggled to assimilate the early material in their lectures and get more and more lost. Or possibly, their attendance is patchy. In an investigative course, attendance is compulsory, and the training nature of the activities means that all students *should* reach a more or less common level. In other words, the compression of marks is not an anomaly: it is a real effect because all students reach a similar (high) level of competence.

Project work is discussed in detail above. Here I simply reiterate some of the main points about its assessment:

- It is normally better for projects to be assigned to pairs or larger groups of students. Everyone needs to discuss things and it is better for the discussions to be with peers than for a student simply to be told solutions.
- Projects must be set at a level such that students are competent to make decisions.

- So far as possible, the marking should be carried out by someone who was not the supervisor of the project and by as small a group as possible to enhance consistency.
- The one area where the supervisor must be at least involved is in assessing the performance of the students, which should be distinguished from the other elements such as written papers, oral presentations, posters etc

Most of the elements of assessment of project work involve some sort of grid of the type we have seen *ad nauseam* above. I re-emphasise the point that these grids are valuable for defining what features both students and markers should prioritise but that they most certainly do not define an absolute standard. I shall examine how we might take into account the reality of norm referencing in chapter 6.

Finally, I make a comment on disabled students. The Institute of Physics has provided some excellent guidance for assisting university departments in ensuring that disabled students are able to participate as fully as possible in investigative work [6]. Usually, relatively straightforward adjustments can be made to include students with most disabilities, including those in wheelchairs, deafness, autism and so on. In those cases, assessment processes will only require minor, if any, adjustment. Occasionally, a student may have a disability that has a more severe effect on their ability to acquire some investigative skills. For example, a number of blind students have graduated from physics programmes [7]. In these cases, although the students were given some experience of the scientific method, adjustments were made at the programme level to ensure that the students were able to compensate the loss of some learning outcomes with the acquisition of others, for example in computing.

Bibliography

[1] Shephard K, Trotman T, Furnari M and Löfström E 2015 Teaching research integrity in higher education: policy and strategy *J. High. Educ. Policy Manag.* **37** 615–32

[2] Goodstein D 2010 *On Fact and Fraud: Cautionary Tales from the Front Lines of Science* (Princeton, NJ: Princeton University Press)

[3] The Open University and The Wolfson Foundation 2013 *The OpenScience Laboratory* https://learn5.open.ac.uk/course/view.php?id=2

[4] Virtual Physical Laboratory http://vplab.ndo.co.uk/

[5] www.seminarsonly.com College Physics Project Ideas Experiments www.seminarsonly.com/Engineering-Projects/Physics/College-Physics-Project-Ideas-Experiments.php

[6] IOP 2008 Access for all: a guide to disability good practice for university physics departments *Institute of Physics Guide* www.iop.org/sites/default/files/2021-03/access-for-all-disability-good-practice-university-2008.pdf

[7] Cartwright J 2018 Physics in the dark *Phys. World* January

IOP Publishing

Assessment in University Physics Education

Peter Main

Chapter 6

Summaries and suggestions

6.1 Introduction

In the previous chapters I have tried to analyse the various types of assessment used in physics programmes, with an emphasis on their link to the purposes of assessment and how they reflect the desired learning outcomes of the programme. I have considered their effect on student behaviour, that is, the type of student activity that would result in high marks in the particular assessment. There is also advice on how to avoid some of the more prominent pitfalls for teachers carrying out assessments, particularly teachers who are relative novices.

The aim of this chapter is two-fold. First, I collate some of the major issues that have emerged from the analysis and, second, I try to put forward some suggestions for how we might start to address these issues. It is not easy; as we shall see, some of the issues that arise occur as a direct result of tensions between the different purposes of assessment, tensions that are not resolved easily without a substantial change of mindset. Some of the suggestions will be relatively mild and can be readily implemented; others are more radical and are intended to provoke thought and debate.

It is useful to revisit the purposes of assessment, as outlined in chapter 1, or more accurately in the case of purpose 2 below, how assessment is used. These are:
1. Classifying students: grading.
2. External reputation: quality assurance.
3. Promoting learning.
4. Measuring learning.
5. Benchmarking.

In chapter 1 I merged items 3 and 4. In fact, items 3, 4 and 5 are in some sense coherent. We want our graduates to have knowledge and skills and we want them to be benchmarked as competent scientists. It follows that, to be assured that they do indeed have these virtues, we need to assess them. Note that item 4, 'measuring learning', is as important for the teachers as for the students, as it indicates how well we are doing our job as educators. In other words, once we have decided what skills and knowledge we want our students to acquire in the programme, we need to know whether we have been successful in helping them do so.

Items 1 and 2, however, do lead to tensions. The government, employers, admission offices for postgraduate programmes in universities, and grant awarding agencies, such as the UK research councils, all use grades as a measure of the quality of the graduate. Understandably, students often see their grades as the desirable outcome of their degree programme, rather than the quality of the education they receive. Similarly, external, so-called quality assurance procedures, whether formal, such as the Teaching Excellence Framework, or informal, such as the newspaper-based or online league tables, often reward universities that produce high ratios of first-class degrees and penalise those that have a more balanced distribution of degree awards or have high drop-out rates. While at one level this is not unreasonable—no one wants students to fail—the logic is based on all institutions having similar standards, which is not the case, and the net result, as described in chapter 1, has been substantial grade inflation. In many universities the vast majority of students are awarded 1st class or 2(i) degrees so that students see anything less than that as akin to failure.

At the time of writing, universities around the world are still being heavily affected by the Covid pandemic. In the UK, over the last couple of years the restriction on putting large numbers of students into close proximity has led to a substantial revision of assessments, including a move to open book examinations, often with generously extended durations. In addition, many universities have put in place a safety net, so that students who do badly in a Covid-affected assessment either have another chance or can fall back on earlier marks. Such measures are entirely understandable and, in terms of grading (as opposed to education), recent graduates are unlikely to have been seriously affected by the pandemic. The more interesting question is whether the changes made to teaching, learning and assessment will continue once Covid is under control. There is certainly an opportunity to instigate positive change and I hope that some of the suggestions that follow might help the thinking behind that process.

6.2 Emergent issues

In the discussions within the earlier chapters a number of issues around assessment have emerged. These are listed below, with some commentary as to why I consider there is a point to be addressed. This list then forms the basis for the suggestions for possible improvement in the following section.

1. *Matching to learning outcomes.*

    The way to design a programme in any subject is to begin with the learning outcomes, simply what you want the students to know and to be able to do when they graduate. Once those outcomes have been identified, the next step should be to decide which assessments will allow students to demonstrate that they have been achieved. Having looked at the assessment regime in many UK universities, I do not believe that in general this is happening. Rather, the assessments are taken as given and most teachers/assessors still see their job as transmitting information to be regurgitated in timed examinations.

2. *Rote learning.*

The analysis of the timed-examination papers in chapter 4 demonstrated that a student can achieve first-class marks simply by remembering and reproducing material, much of which is provided to them in the form of lecture notes. While it is true that it is much easier to remember stuff when one has a good understanding of it, so there is some incentive to keep on top of the information throughout the year, nonetheless students realise that their best strategy for success is to rote learn their notes, a process usually attempted as close to the examination as possible. Further, even when there are questions that involve the application of knowledge, they are usually very similar to questions that students have met previously. Indeed, it can hardly be otherwise; in the environment of a time-limited test—and most of the papers analysed in chapter 4 required candidates to work quickly—it is unreasonable to ask students to think deeply about a new problem because they simply do not have time to do so. It follows that, even where problems are included in an examination paper, it is still a student's best strategy to rote learn the questions set during the course, along with the model answers inevitably provided. Timed examinations form a major part of the degree assessment in the UK and elsewhere and the activity they drive is rote learning, an activity not usually included as one of the learning outcomes of the programme. This mismatch is perhaps the most serious issue in the assessment of science programmes.

3. *Assessment not integrated into activity.*

This point is linked to the one above. With the exception of investigative work, in many cases assessment is not integrated into the day-to-day student activity in a manner that either drives student learning or provides sufficient feedback on their progress. A tension here is between formative assessment, which does not contribute to the final mark, and summative assessment, which does and is, therefore, of higher stakes. In the former case, students are often equivocal about undertaking tasks that do not yield marks directly. Even if they are conscientious about doing work that 'does not count', they are not usually trained in how to use the subsequent feedback [1] to improve their work.

4. *Inconsistency.*

When the principal assessment method is a timed examination, consistency of the marking of that paper should be guaranteed, since the scripts are usually marked by one person against a model answer. But that does not mean that the standard is the same across all papers, however well-set and moderated they are. Over decades of involvement with university assessments, I can barely recall a single suite of examinations where at least one set of marks was not out of kilter with others for the same cohort. These papers are usually thought of as anomalous but, in fact, they are just an inevitable consequence of the fact that exam setting is an imprecise human activity. Why do we consider marks sacred when we know there is such variation?

As discussed in chapter 5, inconsistency of assessment is even more of an issue with project work, where not only are students undertaking very different tasks, but they are also being marked by different people, who, in some cases, have been closely involved in the progress of the project. In Masters programmes, including the integrated MPhys/MSci versions prevalent in the UK, the final project contributes a very large fraction of the final-year marks. It is concerning that the item that has most influence on degree outcome should be the one where there is most unease about consistency of marking.

An additional point about project work is that marks are more commonly awarded for proxies, such as written and oral presentations or even basic knowledge of physics, than for the actual work done by the students in carrying out the project. Indeed, the daily activity can vary enormously for different projects, particularly with regard to the degree to which students are able to make decisions and the assistance they might receive from postgraduate students and post-docs.

5. *Lack of synoptic assessment.*

A defining characteristic of a physicist is the ability to bring together understanding from across the subject to understand the world and to solve problems. Where programmes are modularised, as they are in most UK universities, the basic idea is that each module is self-contained, defined by its subject matter, and that assessments should be confined to material covered within it. Although it is certainly true that later courses build upon earlier ones— for example a course on atomic physics will necessarily assume a knowledge of electromagnetism and basic quantum mechanics—it is not usually the case that the connectedness between different areas of physics forms much part of assessment. But physics does require synoptic thinking and does not fit neatly into a modular system, at least not as well as subjects such as history, literature and even mathematics do.

One common riposte to the accusation that synoptic work is absent is that all students do projects, where there is no defined content. While this is true, in most cases the specific project is narrowly defined and, in any case, the students are not trained in synoptic thinking by their earlier experiences of assessment.

Another approach used to overcome this issue is the 'General paper', a timed examination in which students are asked questions that can be taken from any area of physics—we saw an example in chapter 4. Such papers used to be common pre-modularisation but were mostly dropped following its introduction. But is a timed examination really a suitable vehicle for testing the ability to apply synoptic thinking?

6. *Grading issues.*

In the UK, there are the allegedly well-understood classes of degrees: 1st, 2(i), 2(ii), 3rd, and Pass. I say 'allegedly' because, although one can argue that there is some continuity in the award of degrees in a particular programme in a particular department from year to year, there is no evidence I am aware of that these classifications have any comparability between different universities and/or different subjects. Indeed, in the latter case, it is not even clear what such a comparison would entail. Yet these classes are used for all sorts of purposes *as if they were directly comparable.* Everyone knows that this is nonsense but university leaders are afraid of what might happen if they say so in public.

To exacerbate matters further, the use of failure rates and degree outcomes in some sense as a measure of the quality of the education has undoubtedly resulted in massive grade inflation and possibly a lowering of standards to prevent failures. Despite the longevity of the grading system, its arbitrariness in drawing lines between classes and its lack of consistency of standards mean that the system is palpably broken.

7. *Sanctity of marks.*

The overwhelming majority of assessments in physics involve the award of a numerical mark while the overall mark for the course or module is determined by linear addition of the marks for each component. I am not aware of any logical argument as to why this should be, beyond that it has always been that way, but there are a number of reasons why it is a dubious procedure.

Even in the most consistent type of assessment, the timed examination, there is always a wide variation in the average marks for the different papers. I have heard academics who say that this is because a particular student cohort has (collectively) a greater aptitude for some areas of physics than others. This is wishful thinking and I am not aware of any evidence to support that view. Marks are awarded according to how well an answer matches the model answer. While that is a reasonably objective exercise, papers can vary in difficulty from year to year and assessors differ in the generosity of their marking. In practice, the best one can say is that the marks for a paper ranks the candidates in order of achievement. The absolute level of the marks, however, is not robust, particularly when aggregating marks from different modules.

When the different module marks have roughly the same average mark, there is little harm done in a simple linear addition. But when one or more module assessments are anomalously low or high, the prospect of scaling emerges. Many universities do have policies on scaling (a particularly comprehensive example can be found in [2]) but they are rarely subject specific and I have known individual departments deal with anomalies in an ad hoc manner, despite such anomalies emerging almost annually. In my experience, departments are prepared to raise the marks for a low average module but rarely willing to lower them if the average is very high. The reasoning here is that students are unlikely to complain if marks are raised but will be angry if they are lowered. Given that in many institutions, students are allowed to examine their marked papers, this is an understandable fear.

Nonetheless, there are strong fairness arguments in favour of scaling, which are discussed below. In addition, given that the absolute level of the marks is, despite our best efforts, arbitrary, at best we can argue that standards in a given programme are similar from year to year. Also, if one allows scaling *only* if an average mark is low, one is raising the class average with little justification.

Another argument against a linear addition of marks is that some skills may be suited to threshold assessment; i.e. credit is achieved when a student is able to do something competently. There are many examples of this in a laboratory context, including health and safety competence, but one can also apply it in the context of mastery learning, where a student progresses by achieving mastery at successive levels. In such arrangements, achievement is measured by the degree of progression rather than by the marks accumulated.

Finally, the perceived sanctity of marks is revealed with the reverence with which they are treated in examiners' meetings. Any grading system will have arbitrary boundaries between grades. Sometimes examiners soothe themselves by allowing a margin of error so that, instead of requiring 70% for a first-class degree, 69% is acceptable. But, of course, all that does is to shift the boundary to 69% (or even 68.45% if one accepts rounding!) Such arrangements certainly make examiners feel better, but they make little difference to the principles involved.

8. *Norm versus criterion referencing.*

I hope that I have persuaded the reader that, in absolute terms, and except in a few special cases, assessment is norm-referenced, although almost all institutions claim that they use criterion referencing. A well-constructed assessment system can rank students relative to each other but what it cannot do is to compare them with students from other disciplines or in the same subject from other institutions. This statement is true whether we are talking about timed examinations or the type of assessment grids exemplified in chapter 5.

This pretence of criterion-referenced assessment has led to much confusion and misunderstanding as to how higher education works. In the UK, HE institutions have (correctly in my view) resisted successive governments' attempts to get them to compete on fees but they have done so by pretending that standards and learning outcomes are equivalent across the board, which everyone knows is not the case.

9. *Staff workload.*

I am aware that there is considerable resistance to innovations in teaching in universities, including innovation in assessment. Some of this opposition is due to a reluctance to change the *status quo*, however flawed that might be, or a prioritisation of research over teaching. However, more reasonably, some of it is fear that new methods mean more work. Timed examinations are easy to set, easy to mark and produce a spread of marks. Project-based learning or any sort of continual assessment is generally more labour intensive over a longer period, or at least is perceived to be so. Any changes in assessment do require buy-in from staff; anything that results in a substantial increase in workload is unlikely to be acceptable.

10. *Student view.*

There have been many changes in HE since I entered university in 1970, but perhaps one of the most far reaching has been the influence of the student voice. I have already mentioned the role of the National Student Survey (NSS) in terms of its contribution to quality assurance, but there are many other ways in which the student voice has changed the education process. Students have access to all their marks and, in many institutions, also to their marked assessments. Lecture notes are available to students as are answers to all problem sheets. Departments are required to have student–staff committees to deal with issues and it is the norm to involve students in curriculum development and discussions around assessment.

Most of these changes are very positive—in my day, the only indication of performance we received was the degree result. However, they have also had a restraining effect on teaching and assessment, the origin of which goes right back to the purposes of assessment. A significant fraction of students, though far from all, are more interested in their degree classification than in their education. Innovations in teaching and assessment are subject to conservative pressure. Innovative teaching usually requires consistent attendance and more active student participation than do traditional lectures. Similarly, novel assessment often takes students out of their comfort zone of rote learning. As a result, student satisfaction, as reflected in survey responses, may be low for the very activities that improve their education.

6.3 Suggestions for improvement

I wondered whether to call this section 'Suggestions for change' but decided not to, since there is no point to change if it does not make things better. The suggestions I propose below are intended to provoke discussion, both within and between institutions. I doubt if anyone would adopt them wholesale and, in any case, not all the recommendations lie within the capability of a single institution.

Throughout the earlier chapters, I have made reference to the various purposes of assessment but there is also an overarching reason for education, not at the level of the individual, who might see the purpose as providing the route to a good job, but in terms of the nation; why does any country provide an educational system? In that context, I propose that the purpose of education is to create a skilled workforce and a population capable of rational thought and empathy with their fellow beings. I accept that not everyone will align with those purposes but most people would agree that it is the education—what graduates know and can do—that is important and not the grades in themselves. It follows that the *sine qua non* of any assessment system is to drive student learning. The other purposes of assessment are relevant to the context of higher education in our society but are not strictly essential. Unfortunately, those inessential purposes are currently obscuring the most important one and the thinking behind my suggestions is to try to redress that balance.

As I write, the world is coming towards the end of its second year of dealing with Covid 19. There is little positive to say about the effects of the pandemic; higher education has been severely affected both in terms of the educational process and assessment. While we all hope that Covid will soon become an unpleasant memory, there are a couple of legacy opportunities that should not be wasted. First, a great deal of online material has been developed and it would be a huge waste if that were discarded; it represents a valuable resource that can free departments from the straightjacket of traditional lectures. Second, because it was not possible to co-locate large numbers of students in examination halls, assessments had to be modified accordingly. Although many of these adjustments were necessarily *ad hoc*, due to the timing of the pandemic lockdowns, there is a real opportunity now to think through assessment from scratch. It would be a pity to lose that opportunity.

6.3.1 Quality assurance and assessment

As I have described in detail, external and internal measures of quality have had a profound effect on assessment in universities. The principal metrics employed are: student opinion; student performance (grades and failure rates); employability; and resources per student, including staff–student ratios. None of these is a direct measure of the quality of education and each can lead to perverse incentives, such as the lowering of standards. Even employability, which seems a reasonable enough measure, links more to the perceived status of an institution than the quality of the education provided: employers do not shortlist recently graduated candidates by assessing the quality of their education; they look at the name of the university and the grade.

The basic problem is the assumption that all university programmes are trying to do the same thing and that standards are the same in each institution. Even if we restrict the comparison to programmes in physics, a subject that probably has more uniformity across the sector than most, this assumption is not valid. To exacerbate the issue, there is frequently confusion between the terms 'quality' and 'standards'. Quality generally refers to the facilities and procedures etc, for example the way that the student voice is encouraged and acted upon or how new courses are developed. Standards, in contrast, refer to the level to which someone has been educated. Almost all the external (and internal) reviews of programmes concentrate on the quality aspects although the outcomes of those reviews are often taken as a measure of standards.

I propose a three-fold solution to the current arrangement, which would allow the various institutions to maintain their identities, provide genuinely useful information to inform student choice and preserve some element of competition and accreditation to satisfy the government and its agencies. The first suggestion is that programmes are measured against a statement of what they are aiming to achieve and that they are obliged to provide robust evidence to support their success.

For example, if a department claims to attract the very best students and that their graduates all get top jobs, that will appeal to prospective students, but if their intake does not live up to the statement, or their graduates fail to find good, graduate employment, they will receive a low grade.

When I have aired this suggestion, a common response is a fear that programmes might have deliberately low ambitions in order to achieve a good rating. However, since both the aims and outcomes would be reported together, that would mean prospective students would see the lack of ambition in the aims and be able to take that into account in making their choices. It is beyond the scope of this book, and outside its spirit, to go into further detail about what such a scheme would look like but there would be many advantages over the current arrangements, which suit no one. To take one important example, the current arrangement discriminates against universities taking risks by admitting students with potential but with no proven academic record because such a policy is likely to lead to an above average failure rate. In the proposed system, that would be fine because the university could include the risk-taking as one of its aims.

The second part of the recommendation refers directly to the quality aspects, the physical and administrative infrastructures that underpin the educational process. These include such elements as peer observation of teaching, student–staff interactions, complaints procedures and so on, as well as computing and library facilities. I do not wish, in any way, to underestimate the importance of such arrangements; a poorly run programme can be a nightmare for students. However, the infrastructures normally apply to the institution rather than to individual programmes. Consequently, any assessment of quality should be at that level. As with my first suggestion, this is not the place to go into much detail but one could have a four-level grading of outstanding, satisfactory, unsatisfactory and critical, the last included to allow immediate action.

This suggestion would mean that quality procedures must include a robust, effective and transparent process for seeking student input and for dealing with student complaints. However, student ratings should not be used directly for comparison between programmes in different institutions, as they currently are in the NSS. There are many reasons for this but here are three of the better ones:

- Students do not have experience of more than one programme so they cannot make comparative judgements. Also, experience shows that numerical ratings can be skewed by a relatively minor personal experience, positive or negative, which may not even be associated with the programme at all.
- A relatively large graduating class of, say, 150 will generally have around 80 respondents. Such a small response will inevitably lead to statistical fluctuations in the numerical ratings, typically around 5%–10%. As a Head of Department, on several occasions I have been praised for an increase in reported student satisfaction or asked to improve procedures following a decrease; in fact, the changes were entirely within the range of the expected statistical variation.
- With the NSS staff are not allowed to prime their students on how to complete the survey, for example by telling them that a neutral rating implies unsatisfactory or, more heinous still, implying that a poor rating would

devalue their degree. However, although it is not strictly within the rules, institutions can prepare students for the survey, for example by conducting mock surveys in earlier years (only graduating students complete the NSS).

The third and final part of the recommendation is that professional bodies should be formally part of the quality mechanism. In the UK, we are fortunate in physics that the Institute of Physics has a well-established degree accreditation scheme [3] which is owned by the physics community in the sense that peer representatives decide the criteria and carry out the assessments. The role of this accreditation is to ensure a basic threshold standard of provision, including some judgement about the level of the material covered, and to provide a secondary perspective on quality assurance mechanisms. The reports of these external, accrediting bodies then play an important role in the internal quality assurance procedures as described above.

If these suggestions were implemented, most of the pressure on assessment due to external evaluation would be removed. Indeed, we could revisit the purposes of assessment and remove the entry on quality assurance altogether. Taken together with the removal of the arbitrary degree classifications, as discussed below, that would allow universities to think much more clearly about what they are trying to achieve rather than conform to misleading performance indicators.

6.3.2 Norm versus criterion referencing

I hope that I have convinced the reader that, although most universities claim to have degree programmes where the marking is criterion-based, in practice, almost all assessments are norm-referenced; students compete with each other, with a nod towards students in earlier cohorts. In traditional examinations, examiners set questions that their students can answer, based on their experience of previous cohorts; indeed, it is often the case that the setters are recommended to set and mark a paper with a target average of, say, 65%. It is noticeable that when a new member of staff sets their first paper, without that experience of earlier cohorts, it is not uncommon for them to pitch the paper at a level that results in an 'anomalous' average mark, usually too low.

In the more qualitative assessments, a grid of the type we saw in chapter 5 is the usual way of grading students; the analysis of such grids points unequivocally to them leading to norm-referenced marking. They have great value in indicating to students what parameters matter in terms of assessment but are of limited use in defining an absolute standard. That is not to say that there is no element of using objective criteria—it would be patently absurd to give high marks for grammar to a student who could not use verbs, say. But, as we have seen, the inclusion of such differentiating terms as good, very good, excellent, outstanding, exceptional does not create a template for objective judgement.

Could we move to a criterion-based assessment system and, if so, would it be desirable? I think the answers are: possibly and no, in that order. One area where it would be possible and may even be desirable is with a threshold assessment. In chapter 5 I raised the possibility of such assessments in the context of laboratory work, for example health and safety, where the students either demonstrate a skill or they do not. Similarly, a mastery-learning approach to teaching requires students to achieve high marks in a test before moving on to the next stage of the course. Mathematical topics lend themselves to this approach, because a student may have to become fluent in a technique before moving to more advanced material.

The mastery-learning technique involves criterion-based assessment by its very nature. However, such courses are very rare and would not easily fit into the current model for aggregating marks. I will return to that point later in the chapter but for the moment I consider whether one could have criterion-based, traditional examinations. And the answer is yes, provided you would be prepared to define everything in minute detail; and by 'everything', I mean the content, the level, the degree of scaffolding in the questions, the degree of problem solving etc. Even if one were able to do this, the objectivity would last only for a year or two, until there was an archive of old papers and students would begin to guess the questions.

In parenthesis, I have heard serious suggestions that one should give the students, say, 100 questions and tell them that some small subset will appear in the examination. I will leave the reader to provide her or his own critique of that idea.

I am not aware of any physics department that has taken this detailed approach to setting examinations. It would require an extraordinary amount of work to do so and for little gain. In addition, in terms of comparability of assessment, it would only make sense for them to carry out this task if other departments were to do the same and that is unlikely to happen, to say the least. To take this idea to its extreme, it would be possible for there to be a set of national physics examinations, which would certainly solve the problem of comparability. However, that would require every department to teach the same curriculum, which would remove the rich diversity of provision. Not every department is trying to do the same thing. The idea is both unworkable and undesirable.

My suggestion is that examiners recognise that assessments are designed and marked in a way that is more consistent with measuring the relative performance of students than as an absolute measure that can be compared with other institutions. Once one takes that step, a department can then link the assessments with the vision of the programme, the learning outcomes to be achieved, rather than having in mind some spurious definition of what constitutes a 'first-class' student, to be compared with the best students from other universities.

Naturally, such a move would require a change in way that degrees are graded. The fact that the categories of 1st class etc are used across the board itself implies a commonality of standards, which does not bear scrutiny and does not reflect the reality that different programmes are trying to do different things. These matters will be discussed further in section 6.3.5 below, where I also bring in the possibility of having a combination of criterion- and norm-referenced assessment *provided that* marks are not considered sacred, as discussed above.

6.3.3 Designing assessment

Much of this book has been written in terms of looking at ways of assessing an existing physics programme, which in the UK is usually packaged into notionally self-contained modules. However, from time to time, usually every five years or so, programmes are reviewed and, occasionally, there is a radical overhaul. In principle, such reviews present a rare opportunity to design assessments to achieve what you want them to achieve but, in practice, the assessments usually come as an afterthought, if they are considered explicitly at all. A typical approach would be to start by deciding the content and how it appears over the three or four years of the programme and then create courses/modules with names such as 'Mechanics', 'Electromagnetism', 'Nuclear Physics' and so on. Laboratory work, computing and a selection of options are considered separately. I confess that I have been involved in several such processes in my own institutions and as an external adviser.

There are a number of issues with this approach. The two most serious are: first, that it splits the coherent discipline of physics into a fairly arbitrary set of packages, each of which is seen as an independent entity with its own specific assessment, and, second, because the curriculum is content driven, there is little connection to the full set of learning outcomes (LOs).

To take the coherence issue first—of all the subjects taught in universities, modularisation is least kind to physics. I can appreciate that in subjects such as, say, history, once one is familiar with the techniques of the subject one can apply them to different eras and from a variety of perspectives, although I would still argue that some connectivity is lost by having stand-alone modules. However, in the science disciplines the whole point is that subject is based on a coherent core of knowledge. Of course, one can have options that go into more detail in some specific application or extension of core material, but around 40%–50% of the material offered in physics programmes is the same across the sector. The second issue of matching to the LOs is the more important of the two because, if we can get that right, the compartmentalisation of physics is less serious.

The first step in designing a programme is to decide what its LOs are. If that nomenclature upsets you, then just ask the question: what do I want the students to be able to do when they graduate? Although I provided some examples of LOs in earlier chapters, this is a book about assessment and not programme design, so it would be a diversion to provide a specimen set. However, they do need to be detailed, not just vague notions about understanding physics (an LO almost impossible to measure anyway).

Perhaps irritated by the detail of the paperwork required to set up a new course, I have known many people remark that 'they just need to know some physics', which I guess is similar to saying 'what was good enough for me is good enough for them'. Having been head of a department lucky enough to have staff from around a dozen different countries, I am aware that the definition of 'what was good enough for me' varied wildly between those staff! More seriously, the identity of any physics programme is defined by its LOs and these various identities are right at the heart of my suggestion above, that a programme should be judged by what it is trying to achieve, not some spurious comparison with other programmes with entirely different visions.

Included in the LOs, or as a separate item, will be the content and the level to which it is taken. Tradition usually means that this process is done by identifying strands of learning, such as mathematics, quantum and atomic physics, thermal and statistical physics, electromagnetism and so on. Even if that is the case, it is best to try to end up with a single curriculum rather than a set of parallel, semi-independent strands. Another important point to bear in mind is that content should only be included if its assessment is linked to an LO. On numerous occasions in my duties as an examiner, I have come across an examination that requires 100% rote learning. When challenged, the examiner inevitably replies that the material is so advanced that it is impossible to ask any question in the examination format that requires students to think independently. My response is that either the level is too advanced or, more constructively, that the examination format is not appropriate for assessing that material.

Having determined the detailed, programme-level LOs, the next step is to define the assessments that will enable you to decide if the students have achieved them. This stage occurs *before* any packaging into modules, although some of the LOs might be specific to certain content. This is the point at which issues such as teamwork, synopticity etc, can be addressed. If you want your students to be successful team workers, you have to decide, first, how are you going to assess that virtue, not retrofit it into an existing set of modules. Similarly, if you want your graduates to be able to draw on different elements of physics to address problems, then decide how that will be measured.

Armed with the LOs and notions of how to assess them, the next step is to put them into packages, while retaining the overall coherence. If your university does not operate a modular system, you have much more flexibility than if you are in one that does. In the latter case, where the packages usually have to be in units of 10 or 15 credits (where 120 credits constitute a whole year), it is wise to use blocks as large as you are allowed, say 30 or 40 credits. In Year 1, for example, a large module on, say, classical physics, can bring together all sorts of different types of assessment, including regular course work, computing exercises, mini-projects, multiple choice tests and timed examinations. The breadth allowed by the packaging also facilitates synoptic assessments. In short, the larger the package, the more flexibility there is about assessment.

In principle, one could incorporate mathematical methods and laboratory work into the integrated modules but I think that may not be wise in Year 1, where techniques have to be mastered.

Another advantage of planning programmes in this sequence, LOs–assessment–modules, is that one can match a form of assessment to a topic. To take a specific example, it is usual to introduce quantum mechanics in Year 2 or 3 since the topic requires expertise in partial differential equations and matrices. A mastery learning approach would be ideal in this case, and the mathematics could be integrated into the sequence of learning.

6.3.4 Variety of assessments

The previous section outlined how assessment can help drive the design of a programme. I am aware that my proposal is radical and would be seen by some as too idealistic. One significant barrier (others are discussed below) is that most academics *begin* with the idea of content-defined teaching with an expectation of a timed examination at the end of it. Consequently, such examinations still form a large fraction of the assessment regime in most UK physics departments. However, we have seen that such examinations have many faults, including the following important ones: they do not lend themselves to providing feedback (until it is too late); they are high stakes; and, crucially, their main effect on student activity is to promote last-minute, rote learning.

Can timed examinations play a role in an assessment regime that is linked to learning outcomes? Yes, I think they can, but some thought needs to go into how they link to the LOs and how they will drive student behaviour. Examples of where they might be useful are:
- As part of a mastery-learning course.
- To demonstrate threshold competence in mathematical manipulations.
- To demonstrate knowledge where such knowledge is necessary for further progress.
- To demonstrate the intuitive and qualitative application of ideas.
- To provide feedback to students and staff.

The purpose mentioned in the penultimate bullet point may be relevant for the early stages of a programme or course. The concept inventories [4] provide multiple choice questions that do this job. The implication of the final bullet point is that there is not much use in having the examination or test after all the teaching has been completed. I recognise that this point does have administrative consequences but if the only summative assessment for a course or module is a timed examination at the end, it is very difficult to see how we can move away from the emphasis on rote learning.

The many alternatives to timed examinations were presented in chapter 2, along with their strengths and limitations. Provided assessment is an integral part of programme design, there should be no difficulty in creating a suite of assessments that link to the LOs, that promote active student participation and that allow feedback to be an integral part of teaching and learning.

I started this section by admitting that my ideas are radical and that there are significant barriers to introducing them. Here are a few I have thought of myself:
- Universities tend to be geared to timed examinations being the primary form of assessment and have scheduled examination and revision periods. This is a serious issue, which, in the long term, can only be solved by the institution reviewing its assessment procedures. However, one can also have local solutions. As long ago as the 1970s, during my own degree programme at the University of Birmingham, the department arranged for the focus of the final year to be a 'Group Studies' [5] activity involving the whole cohort of final-year physics students and this took place during the main university examination period. Provided arguments are made constructively, universities will usually be receptive to innovative assessment.
- A related point is that, if timed examinations and tests are to take place outside the main examination periods, there will be problems in arranging them, particularly if they require the usual degree of security and formality.

However, because these examinations would form a much lower fraction of the total assessments, they can usually be administered as tests. Again, a long-term solution will require institutional help but in many modular programmes there are two examination periods, one after each semester, and one can use those creatively.

- In recent years there has been a rise in the number of mental health issues reported by students. Whether this is a genuine increase or that there is, happily, much less stigma about reporting such issues is not really relevant. As a result, many universities have been trying to decrease the number of assessments in an attempt to reduce stress on students. Personally, I find the logic puzzling: having fewer assessments means that the ones that remain have higher stakes, which is likely to cause a great deal of stress. Having more means that each one is relatively less important and, in any case, students are participating more in their learning. Where there can be an issue is if, for some reason, a student struggles to attend regularly but such issues are present in current assessment schemes too, albeit at a lower level.
- Students expect to have timed examinations at the end of the module or course. This is how the system operated at school and it will be what they expect at university. Also, their friends and relatives will have told them the same thing. Against that, when students arrive at university, they are generally full of enthusiasm and there is an opportunity to do something different with them. Once the students slip into a weekly lecture and labs routine, it is much harder. The lesson is that, if one does wish to be creative about teaching and assessment, it is best to do it from the start. Or, if not then, at a major transition point, such as the start of a new year.
- A related barrier is that students can get into a cycle of poor attendance. At many UK universities, teachers are obliged to provide lecture notes for their courses and, often, lectures are recorded so that students can view them later, a process usually referred to as lecture capture. Perhaps unsurprisingly, lecture attendances can be dismally low, well below 50%. If the teaching session is basically information transfer and that information is already available without attendance, I am sure staying in bed on a winter morning is a more appealing prospect than a lecture. An assessment and teaching regime that requires students to attend will clearly run into problems if they do not. But my experience is that, provided the attendance requirement is made clear to them from the beginning, they will accept the situation. I should say that the term lecture capture is also used to describe a flipped classroom method of teaching (for a physics example see [6]) but, in those cases, all lectures are online and the in-person sessions require students to attend.
- A more bureaucratic barrier in some universities is that students cannot be discriminated against for poor attendance at teaching sessions. However, such a rule is absurd in practice-based learning (for example, in medicine) and can be resisted if a good educational case is made.
- One of the major advantages of a system based on traditional, timed examinations is that the load on lecturers is minimised. It takes a day or

two to set a paper and a few days to mark it. And, in between, there are a couple of hours of lectures a week for two or three months. Most staff have the impression that a student-led approach is more labour intensive although, when the hours are calculated, there is usually little difference in the two cases if the student-led learning is well designed. The reason this impression proliferates is that the teaching and assessment tend to be more intensive and requires more thought to develop and apply. Similarly, as described above, students used to standard lectures may balk at being required to attend for, say, 30 h a week. The way to overcome these barriers is to ensure that both staff and students are involved in developing the new approach, particularly new staff who will have new ideas and enthusiasms. There will always be people who believe the *status quo* is the best of all possible worlds but, as I have seen from my own experience, they can be persuaded so long as there are enthusiasts that show how effective the changes can be.

I finish this section with an anecdote based related to the final bullet. In the 1990s, physicists in the UK became concerned that too much new material had been forced into the existing three-year BSc and that, not only were students being overloaded, but there was no room to develop skills, such as communications and teamwork. As a result, new four-year programmes were introduced called 'Master in Physics' (MPhys) or 'Master in Science' (MSci) to run alongside the BSc. In Nottingham we decided that our new fourth year cohort would be entirely student-led, with no formal lectures or traditional examinations. We tried to work closely with the first cohort of final-year students but, even so, after a few weeks, I received a representative group of students who said that the group felt it was having to work too hard. I made the right noises but suggested they wait a little longer to get used to the new style of learning. In truth, there was little we could do in mid-stream. However, a few weeks later, after many of the students had been to job interviews, I saw them again and they had lost all their concerns; the impact they had made at the interviews, explaining all the hands-on work they were doing, had led to a flood of job offers.

To complete the anecdote, the novel approach in the fourth year of our MSci was not universally accepted, with some staff nervous that students would not learn anything without sitting examinations. But I recall clearly one of them, who had been involved in the new teaching despite his reservations, saying that he had never seen students so enthusiastic and such a high quality of work. The moral is that students and staff can be persuaded.

6.3.5 Combining marks

I have said a great deal about how universities claim to award marks according to a criterion-based system although their methodology is closer to norm-referencing. The question is: does it matter? And the answer is: yes, if you continue to insist that all degree classifications in all subjects and universities are comparable. But the answer is no, provided that each programme is clear about what it is trying to do and is judged against that vision. In that case, while the assessments are still norm-referenced, they can be consistently applied within the cohort to provide a quasi-criterion-referenced system. In other words, it is possible to have a near-objective system of assessment that makes sense for each programme and is consistent in historical terms. That is not so far from what we have now within a given programme; it is the insistence on comparable degree outcomes which leads to the issue.

To take norm-referencing to its logical conclusion, one could argue that students should be simply ranked within the class and be done with it. Indeed, many US institutions, when asking for a reference for a potential graduate student, asks explicitly for a class ranking. I do not believe that this would be a fair system of reporting achievement, particularly if that is the only output. Put simply, a student who has achieved well should not have their performance devalued because someone else has done better. This is another argument for the quasi-criterion-referenced system mentioned above.

The current system of awarding 1st class, 2(i) class etc honours degrees contributes to the notion that degree classifications have the same meaning across subjects and institutions. Although one can argue that any common system implies comparability, a new system would not have the historical baggage of the *status quo*. Therefore, some sort of grade point average (GPA) of the type used in many other countries is desirable. In principle, the GPA could be a percentage but that would tempt people to make the comparisons with the current system where, for example, one associates a mark above 70% with a first-class degree. Instead, institutions could report a GPA, together with a class average GPA and some indications of what a GPA at a certain level means in terms of the aims of the programme. I am aware that this suggestion is not ideal but it would, at least, indicate that a GPA was specific to a programme and not generally comparable.

In a modular system, the marks or grades from each module are combined according to their credit weighting to form the final grade. Across the UK, there are various models used; most of them have the later years weighted more heavily than the earlier ones, which is reasonable. In some institutions, failure in one or more modules can be compensated or condoned, which is less reasonable but necessary in examination-based systems where the one-off character of the assessment can lead to a relatively high failure rate, particularly in mathematical subjects.

The issue with the combination of marks is often how to deal with modules with anomalously low or high averages, a frequent occurrence in numerical subjects. Those who believe in the sanctity of marks oppose scaling, presumably believing that the class has had a collective bad/good day. The more likely explanation is that the assessments for the modules varied in difficulty. That matters, because if one leaves anomalous marks as they are, the modules with high averages are being privileged in terms of the overall sum.

Generally, departments are more uncomfortable with low averages than with high ones. They often introduce some sort of scaling for the former, because they rightly see that as the fair thing to do, but rarely for the latter as they are afraid that students will react against marks being scaled. As a result, in the name of preserving the sanctity of marks by not having a formal scaling procedure, the overall average for the class is arbitrarily raised!

Concerns about student complaints are real and understandable in an age where student opinion plays such a role in determining the public image of a department. Nonetheless, scaling is the logical and correct way of dealing with the issue. Once one escapes the idea of the marks having absolute meaning, the solution is quite simple: scale at the modular level so that the marks returned conform to whatever range of acceptability is considered appropriate. To some people, this sound like sophistry but, if you accept my arguments about norm-referencing, it is the obvious way forward. Indeed, another way of thinking about it is not as scaling at all but as a recognition that the primary effect of assessment is to place students in mark order.

Another effect of presenting modular marks within an agreed range is that the range of GPAs (or degree classifications if that system is retained) is more or less fixed, which automatically militates against grade inflation. The only rider to this advice is that one needs to be cautious with option modules, whether they are available to students from within a cohort or include people from other programmes. In the latter case, it is best to ignore the outsiders in setting the overall level of the marks for a module. In the former case, some account may need to be taken of whether the students involved are representative of the class.

6.3.6 Combining marks within modules

I turn now to the modular level. While a weighted average of the module marks/grades to provide a final grade is the correct thing to do, within a particular module, the marking structure may be quite different. In some cases, a linear averaging of the grades from the various components may be appropriate. A small variation might be, for example, to weight the earlier assessments less or to allow some redundancy to allow a student a bad day. But there is absolutely no reason why one cannot be more creative if the assessment is seen in terms of learning outcomes. A couple of examples illustrate the point.

I have mentioned the possibility of threshold assessments, whereby students demonstrate their ability to achieve a certain task. Such arrangements do not lend themselves to the cumulative addition of marks. However, another approach is to say that candidates who pass the thresholds are automatically awarded a pass grade for the module. Other more suitable assessments can be used to introduce discrimination. Although I have discussed threshold assessments mainly in terms of laboratory tasks and other skills, there is no reason why the idea should not be extended to other types of assessment, even examinations. In that case, if there is certain core knowledge or level of mathematical manipulation that is considered essential, a straightforward paper that concentrates solely on those essentials could act as a threshold assessment. Anyone passing that paper at whatever threshold mark is decided receives, as a minimum, a pass in the module. A subsequent paper or other assessment can act to provide more discrimination. Note that the 'threshold' and 'stretch' assessments do not have to be the same form.

The second example refers to teaching via mastery learning [7]. In this approach, students learn at their own pace, moving through stages of learning: they have to demonstrate mastery at one stage before moving to the next, usually via some sort of test.

In this arrangement, marks and grades can be awarded for how far a student progresses, possibly with a threshold level of progress, as described above. Again, one can add a more general assessment to see how well the material has been assimilated.

To summarise this section, module marks or grades can be awarded by using assessments that are matched to the LOs; there does not need to be a linear addition of components. Marks are moderated and possibly scaled, to provide comparability. The overall GPA is determined by a suitable weighted average of the module marks. The traditional classifications of 1st, 2(i) etc are not used but each programme has a set of (positively designed) broad, grade descriptors that link to the stated aims of the programme.

6.3.7 Consistency

The issue at stake here is that each student should be marked to the same criteria. Such consistency is easily achieved in a traditional examination, where there is a detailed model answer, a single marker and a set of anonymous scripts. The task is harder in more qualitative assessments, such as oral and written presentations, which not only have more subjective marking criteria but often involve several independent assessors. Hardest of all are project assessments, which have all the issues associated with qualitative assessment plus the possibility that the project sponsors may feel that a low mark in 'their' project may look as if they were not doing their job well.

The project case was discussed in detail in chapter 5. My most important advice, and probably the most controversial, is that the project sponsors should have little responsibility for the award of marks or grades. Their input should be qualitative and based around questions such as: how much help were the students given (including from post-docs and postgraduate students)?; what decisions did the students make on their own?; how well did the students perform relative to those in previous years?; if there was a group or pair, what about the balance of work between them? I suggest a small mark be allocated to this input, say 10%.

Most projects of substance have multiple assessments, including scientific papers, oral presentations, posters etc. There should certainly be a component concerning the performance of the project, which should be based on a student log, perhaps a laboratory notebook, a portfolio or something similar. For all components, there are two principles to ensure consistency. First, everything should be genuinely double-marked: that is, the markers do not confer and return their marks independently. Allowing discussions between markers prior to mark submission is a practice fraught with issues (different levels of seniority, for example) and should be avoided. The second principle is that, as far as possible a single person should be involved with all the assessments of the same type. This last condition may be hard to implement for large classes. In that case, one person should be responsible for moderating sample assessments.

One tip I offer on the qualitative marking issue is that for each component of assessment, the first step is to rank the students in order. Such an approach is very helpful in the subsequent award of marks and is also consistent with the overarching principle of norm-referencing. If two markers have a completely different rank order, there is a serious issue in the consistency of the assessment.

6.3.8 Instigating change

Many of my suggestions represent major departures from what is currently happening in physics departments. Inevitably, there will be some opposition from staff. There will be those who can see no problem and see no reason for change and others who fear an increase in workload. Students also are resistant to change and are reluctant to move away from traditional examinations, despite the stress they cause.

My experience of instigating change in higher education indicates that there must be a group of enthusiasts who lead the way, plus at least tacit support from senior management, i.e. no active opposition. The role of the innovators is to show that the change is for the better, that it works, that any increase in workload is manageable and to draw in others who show an interest. Then, when the programme review comes along they have to grasp the opportunity to introduce improvements. In three words: innovate, demonstrate, incorporate. When the programme review is underway as many staff as possible should be involved, including the more conservative, and everyone should be invited to challenge perceptions. Physics academics are used to evidence-based discussions. If improvements can be demonstrated, they will respond positively.

The student apprehension is a little harder to appease. One fear they have is of sudden change but that can be dealt with easily by introducing change in Year 1 and rolling through the adjustments as that cohort progresses. In making changes, of course, the student view should be taken into account. Universities set great store in involving students in programme and assessment review but one does need to be cautious: unlike academics, students have only experience of one university. Their input is most helpful in describing how assessment drives student behaviour but they should, nonetheless be fully involved across all aspects of the review. That includes the possibility of their views being challenged as with any other contributor to the review.

It is common in many universities to insist on employer involvement in the design of programmes and, by implication, assessment. A problem for physics, as with many other subjects such as history or mathematics, is that physics graduates enter a large number of employment sectors [8], so it is far from obvious how to choose representative employers; one should avoid the temptation of only selecting employers in the science and engineering sectors. In addition, by far the most popular destination for physics graduates is study towards a PhD. For those reasons, while one should value the advice of employers, it is most useful in defining generic employability skills, including the odd more specific item, such as which computing languages are the most useful. Employers can also offer detailed advice on the best way to arrange such valuable activities as work placements and industrial projects. However, I would caution against asking for too much input on curriculum and

Table 6.1. The principal issues in the assessment of physics programmes and some suggestions on how they might be addressed.

Principal issues	Suggestions for improvement
Matching to LOs	• Design programme around LOs, not only content. • Incorporate assessment into design before packaging programme into modules/courses.
Too much rote learning	• Reduce the number of timed examinations. • Make examinations interpretative rather memory-based.
Assessment not integrated into activity	• Incorporate assessment into programme design. • Link assessment to LOs, then develop module packages. • Ensure feedback is built into assessment and activity.
Inconsistency	• Each assessment has, at worst, one person moderating. • Genuine double marking where appropriate.
Lack of synoptic assessment	• Build synoptic learning into programme design.
Grading issues	• Accept norm-referencing is the reality. • Abandon traditional model of 1st class etc. • Module marks based on target averages.
Sanctity of marks	• Do not consider linear addition of marks as sacrosanct. • Consider threshold exercises and other non-standard techniques that drive learning.
Norm versus criterion referencing	• Accept that the overall standard is a feature of the student cohort and not absolute. • Use criteria for distinguishing achievement within the cohort. • Reject the idea that percentage of high degree awards is a measure of quality of teaching. • Use professional bodies to define threshold standards.
Staff workload	• Use group of enthusiasts to demonstrate effectiveness. • Involve all staff in programme review. • Challenge ideas.
Student view	• Introduce rolling set of changes in Year 1 so no students have a major shift in their assessment regime. • Involve students in programme and assessment review. • Challenge their ideas.

assessment, which are academic matters. Some employers suggest that physics programmes should be more focussed on application, veering towards engineering. That is not good advice.

6.4 Summary

This chapter has tried to identify the issues around assessment of physics programmes that have emerged from the earlier chapters and to offer some suggestions as to how we might improve the process. I am not naïve enough to expect that readers will agree with all the suggestions, let alone that they might all be implemented. However, what I do hope is that academics will be challenged by my arguments and that they will at least think about the issues more deeply than previously.

I leave you with table 6.1, which provides a tabular summary of the main issues and the relevant suggestions.

Bibliography

[1] McConlogue T 2020 *Assessment and Feedback in Higher Education* (London: UCL Press) ch 8
[2] Loughborough University 2021 *University of Loughborough Academic Quality Procedures Handbook* www.lboro.ac.uk/services/registry/pqtp/aqphandbook/12_assessment/
[3] Institute of Physics *The Physics Degree* www.iop.org/education/support-work-higher-education/degree-accreditation-recognition#gref
[4] Laverty J T and Caballero M D 2018 Analysis of the most common concept inventories in physics: what are we assessing? *Phys. Rev. Phys. Educ. Res.* **14** 010123
[5] Black P J, Dyson N A and O'Connor D A 1968 Group studies *Phys. Educ.* **3** 289
[6] Bates S and Galloway R 2012 The inverted classroom in a large enrolment introductory physics course: a case study *The Higher Education Academy* https://www2.ph.ed.ac.uk/~rgallowa/Bates_Galloway.pdf
[7] Keller F S 1968 Good-bye teacher *J. Appl. Behav. Anal.* **1** 79–89
[8] Belfield C, Britton J, Buscha F, Dearden L, Dickson M, van der Erve L, Sibieta L, Vignoles A, Walker I and Yu Z 2018 Undergraduate degrees: labour market returns *Research Report* Institute for Fiscal Studies and the Department for Education https://gov.uk/government/publications/undergraduate-degrees-labour-market-returns

www.ingramcontent.com/pod-product-compliance
Ingram Content Group UK Ltd.
Pitfield, Milton Keynes, MK11 3LW, UK
UKHW051341160426
5217IPUK00047B/122